Health, Wellbeing and Sustainability in the Mediterranean City

T0187679

This book provides a model for the creation of sustainable and healthy cities in the Mediterranean region. It uses the coastal city of L'Alfàs del Pi in Spain as an example for designing renewable and innovative urban models that offer high standards of living, wellbeing and eco-friendly advantages. Quantitative and qualitative analyses are presented by scholars in a wide variety of fields to provide a thorough understanding of the social, cultural, economic, political, physical, environmental and public health influences, through the case study of L'Alfàs del Pi. L'Alfàs del Pi has a geographically unique population made of a mixture of local inhabitants and Northern European residents attracted by the weather conditions and the sea. The chapters in this book explore a series of innovative proposals for addressing concerns in the area, including historic preservation, sustainable transportation, promoting health and physical activity and water conservation. The methodology establishes a strategic approach that serves as a useful reference point for coastal cities, particularly in Mediterranean countries, in the creation of sustainable and healthy cities

This book will appeal to researchers across the disciplines of tourism, planning, health geography, architecture and urban studies.

Antonio Jiménez-Delgado is Professor in Architectural Constructions at the University of Alicante, Spain. He is Director of the AEDIFICATIO International Research Group (BUILDING. Technology, research and development).

Jaime Lloret is Associate Professor in the Department of Communications at the Polytechnic University of Valencia, Spain. He is also the Chair of the Integrated Management Coastal Research Institute (IGIC).

Routledge Studies in Urbanism and the City

This series offers a forum for original and innovative research that engages with key debates and concepts in the field. Titles within the series range from empirical investigations to theoretical engagements, offering international perspectives and multidisciplinary dialogues across the social sciences and humanities, from urban studies, planning, geography, geohumanities, sociology, politics, the arts, cultural studies, philosophy and literature.

Spiritualizing the City
Agency and Resilience of the Urban and Urbanesque Habitat
Edited by Victoria Hegner and Peter Jan Margry

Green Belts
Past; present; future?
John Sturzaker and Ian Mell

Democracy Disconnected
Participation and Governance in a City of the South
Fiona Anciano and Laurence Piper

The City as a Global Political Actor
Edited by Stijn Oosterlynck, Luce Beeckmans, David Bassens, Ben Derudder, Barbara Segaert and Luc Braeckmans

City Branding and Promotion
The Strategic Approach
Waldemar Cudny

Indigenous Rights to the City
Ethnicity and Urban Planning in Bolivia and Ecuador
Philipp Horn

Health, Wellbeing and Sustainability in the Mediterranean City
Interdisciplinary Perspectives
Edited by Antonio Jiménez-Delgado and Jaime Lloret

For more information about this series, please visit: www.routledge.com/series/RSUC

Health, Wellbeing and Sustainability in the Mediterranean City

Interdisciplinary Perspectives

Edited by
Antonio Jiménez-Delgado and
Jaime Lloret

Routledge
Taylor & Francis Group

LONDON AND NEW YORK

First published 2019
by Routledge
2 Park Square, Milton Park, Abingdon, Oxon OX14 4RN

and by Routledge
52 Vanderbilt Avenue, New York, NY 10017

First issued in paperback 2020

Routledge is an imprint of the Taylor & Francis Group, an informa business

British Library Cataloguing-in-Publication Data
A catalogue record for this book is available from the British Library

Library of Congress Cataloging-in-Publication Data
Title: Health, Wellbeing and Sustainability in the Mediterranean City.
Description: New York ; London : Routledge, 2019. | Series: Routledge
    Studies in Urbanism and the City | Includes bibliographical references
    and index.
Identifiers: LCCN 2018045794| ISBN 9781138393752 (hbk : alk. paper) |
    ISBN 9780429401572 (ebk) | ISBN 9780429686238 (mobi/kindle)
Subjects: LCSH: Sustainable development—Mediterranean Region. |
    Urban planning—Mediterranean Region. | Quality of life—
    Mediterranean Region.
Classification: LCC HC244.5.Z9 H43 2019 | DDC 304.209182/2—dc23
LC record available at https://lccn.loc.gov/2018045794

ISBN 13: 978-0-367-66240-0 (pbk)
ISBN 13: 978-1-138-39375-2 (hbk)

Typeset in Times New Roman
by Apex CoVantage, LLC

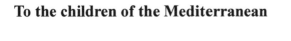
To the children of the Mediterranean

Contents

Figures

Tables

Contributors

Jairo Casares Blanco is a Civil Engineer and Collaborator of Civil Engineering and Urban Planning, Tourism and Transport departments in the University of Alicante, Spain. Since 2014, he has worked on many projects at national and international levels related to transport and railways both for passengers and freights, and for public and private sectors. He has also carried out numerous financial and economic appraisals for Spanish railway projects, advising regional and national governments. He is an expert in Road Safety in urban environments.

Pablo de-Gracia-Soriano graduated in Sociology from the University of Alicante (Special Award of Promotion 2011–2015) and earned his Master's degree in Dynamics of Change in Modern Advanced Societies from the Public University of Navarra (2015–2016). He is currently a Ph.D. student in Sociology and Anthropology at the Complutense University of Madrid, where he is a state University Teacher Training (FPU) Scholar by the Ministry of Science, Innovation and Universities. Since 2016 he has been an Honorary Collaborating Professor for the Department of Sociology I of the University of Alicante. He is a member of the Sociological Observatory of Education (OBSOEDU) and Theory Group of the University of Alicante, having participated in projects related to urban planning, health, tourism, migrations, education, uses and experiences of time and lifestyles. His main line of research is the relationship between society and time, where he focuses his own academic research.

Laura García (laugarg2@teleco.upv.es) was born in Murcia, Murcia (Spain) on December 23, 1992. She received her Bachelor's degree in Telecommunications Technology Engineering from the Polytechnic University of Valencia in 2015 and received her Master's degree in Digital Post Production in 2016. She is working on her Ph.D. Her research lines are focused on e-health, ambient monitoring and quality of life (QoL). She is author or coauthor of several papers in SCI journals. She has been involved in several organisation committees of international conferences since 2016.

Diana Jareño-Ruiz graduated with a Special Award of Promotion in 2006 and earned her Ph.D. in Sociology (Doctorate with International Mention and Special Award) in 2014 from the University of Alicante. She currently teaches classes in the Master Programs in Active Aging and Health and Labor Relations and Human Resources, and the Master's degree program in the Social Innovation and Dynamics of Change, as a professor in the Department of Sociology I of the Faculty of Economic and Business Sciences of the University of Alicante. Her participation in different R + D + i projects and contributions to congresses and publications have an interdisciplinary character (sociology, psychology and anthropology). Her contributions are framed within the areas of migration, family, gender and education. Her research work has allowed her to improve qualitative and quantitative research techniques, having extensive experience in coordinating fieldwork and analysing results.

Antonio Jiménez-Delgado holds a Ph.D. from the University of Alicante in Architecture, City, Civil Works and Its Construction (2008), Technical Architect by the University of Valencia (extraordinary award) in 1991, and graduated in Sociology at the University of Alicante in 2002. He has been a Full Professor since 2000 in the area of knowledge of Architectural Constructions at the University of Alicante and is Director of the AEDIFICATIO International Research Group (BUILDING. Technology, research and development), which includes researchers from the Politecnico di Milano and Hunter College, City University of New York. Formerly the director of the Master and Doctoral Programme in Building Management (2006–2008), he participated in 7 public research projects of competitive nature and 11 research projects contracts. He also coordinated different publications and conferences in the university field.

María Jiménez-Delgado graduated in Educational Psychology and is a Doctor in Sociology, lecturer and director at the Department of Sociology I, and researcher of the University of Alicante in the Institute for Gender Studies Research. Her works focus on gender, equality, immigration, multiculturalism and education. She is director of the research group OBSOEDU (Sociological Observatory for Education) at the University of Alicante. She was principal investigator of the research and development of various projects and she is principal investigator of the research and development project "Muslim Youth Identities: Gender, Education and Citizenship" (GV/2017/169, 2017–2019). She was a co-founder of the Alicante Acoge Foundation, which welcomes and integrates the immigrant population. She works on several educational projects with children's schools, primary schools and secondary schools in the town of Alicante, especially in the northern area, aimed at the educational and social integration of the bridge generation: the children of immigrants.

Ángel T. Lloret received a B.Sc. of Technical Architect in 2008, his M.Sc. in Management of the Building in 2011, and his Ph.D. in 2017, all from the University of Alicante. He is also M.Sc. in Teacher Training, Bachiller and FP (2014) and a Higher Technician in Prevention of Labor Risks (2010). He has been the head of the commercial department of an enterprise and works as technical architect. Ángel has coauthored several papers in national and international conferences and journals (some of them with ISI Thomson Impact Factor).

Jaime Lloret (M'07-SM'10) received his M.Sc. in Physics in 1997, M.Sc. in Electronic Engineering in 2003, and Ph.D. in Telecommunication Engineering in 2006. He is currently an Associate Professor with the Polytechnic University of Valencia. He is the Chair of the Integrated Management Coastal Research Institute and the Head of the Active and Collaborative Techniques and Use of Technologic Resources in the Education (EITACURTE) Innovation Group. He is the Director of the University Diploma Redes y Comunicaciones de Orde-nadores and the University Master Digital Post-Production. He has authored 22 book chapters and has had over 360 research papers published in national and international conferences and international journals (over 140 with ISI Thomson JCR), and has been an Associate Editor of 46 international journals (16 with ISI Thomson Impact Factor). He has been involved in over 320 pro-gramme committees of international conferences and over 130 organisation and steering committees and leads many national and international projects. He is an IARIA Fellow and an ACM Senior, was the Internet Technical Com-mittee Chair of the IEEE Communications Society and the Internet Society from 2013 to 2015, the General Chair (or the Co-Chair) of 39 international workshops and conferences and is currently the Chair of the Working Group of the Standard IEEE 1907.1. He has been a co-editor of 40 conference proceed-ings and a guest editor of several international books and journals. He is an Editor-in-Chief of *Ad Hoc and Sensor Wireless Networks* (with ISI Thomson Impact Factor), the international journal *Network Protocols and Algorithms* and the *International Journal of Multimedia Communications*, and he is the IARIA Journals Board Chair of eight journals.

Carlo Manfredi graduated in Architecture in 1998 and earned a Ph.D. in Pres-ervation of Architectural Heritage in 2008. He was appointed Adjunct Pro-fessor in Architectural Conservation at Politecnico di Milano until 2015 and is now appointed as a Functionary Architect at Museo Storico e Parco del Castello di Miramare in Trieste.

Benjamín Oltra y Martín de los Santos is Professor of Sociology (Sociol-ogy of Culture and Sociological Theory) of the Department of Sociology I of the University of Alicante. He has a degree in Political, Economic and Commercial Sciences, from the Complutense University of Madrid, and a PhD in Economic Sciences (Sociology) from the Autonomous University of Barcelona.

He studied Sociological Methodology at the University of Michigan (Ann Arbor, USA) and the Doctorate program in Sociology and Humanities at Yale University (USA), where he obtained degrees in Master of Arts and Master of Philosophy in Sociology.

He has been Professor of the Autonomous Universities of Madrid and Bar-celona, and Visiting Professor/researcher at the College of Mexico, Berke-ley, Cambridge, Harvard, Le Collège de France and Yale. He was founder, Secretary and Director of *Papers. Revista de Sociología,* of the Autonomous University of Barcelona; and founder and Director of *Campus.* Revista de

Cultura de la Universidad de Alicante. He was founder and vice-president of the "Fundación de Ciencias Sociales y Mundo Mediterráneo".

His hundreds of contributions to the academic field make him one of the preeminent sociologists at the international level. He is currently Professor Emeritus of the University of Alicante, where he continues to work intensely, increasing his prolific contribution with his latest works, among which are: *El Mediterráneo. Culturas, civilizaciones y sociedades, Sociología de la Cultura. Libro de lecturas esenciales, Atlas de pensamiento y ciencias sociales en los tiempos modernos* y, la última edición de *Sociedad, vida y teoría. La Teoría Sociológica desde una perspectiva de Sociología narrativa.*

Armando Ortuño Padilla is a Ph.D. Civil Engineer and Economist. He is an Associate Professor of Urban Planning, Tourism and Transport at the University of Alicante, Spain, project manager at INECA (Institute of Economic Studies of the Province of Alicante), member of the National Committee of Cities, Territory and Culture from the College Civil Engineers and President of the Spatial and Urban Planning Commission from the College Civil Engineers. Since 2004, he has realised multiple types of works, such as studies of demand, impact (income and employment), cost–benefit and environmental analysis of all kinds of infrastructures (transport, hydraulic, energetic, environmental, etc.), location studies, productive activities and equipments, housing demand studies, real estate, tourism and commercial promotions. At present, he directs the project to transform Madrid into a world capital of construction and civil engineering, financed by the town hall of Madrid. He has published numerous books, more than 60 presentations in national and international congresses, and participated in several research projects.

Lorena Parra (loparbo@doctor.upv.es) was born in Gandía, Valencia (Spain) on February 22, 1989. She received her degree in Environmental Science in 2012, her M.Sc. in Environmental Assessment and Monitoring of Marine and Coastal Ecosystems in 2013 and a second M.Sc. in Aquaculture in 2014. She has a Ph.D. focused on integration of new technologies for environmental monitoring specially in underwater environments, obtained in 2018. She is author or coauthor of several papers in SCI journals. She has been involved in several programme and organisation committees of international conferences since 2013.

Rolando Enrique Peñaloza Quintero is a Ph.D. in Political Studies, Master in Political Studies, and Director of the Institute of Public Health of the Pontificia Universidad Javeriana in Bogotá, Colombia. He has more than 25 years of experience in design, implementation and evaluation of public health policies and is a Professor of Public Policy in the Master of Public Health and in the Master of Government of the Territory and Public Management at the Pontificia Universidad Javeriana. He is also a member of the European Evaluation Society.

Raquel Pérez-delHoyo graduated as an Architect from the Polytechnic University of Valencia (Spain) in 1999 and received her Ph.D. in Architecture, City, Civil Works and Its Construction from the University of Alicante (Spain) in 2010. Since 2004, she has been a member of the School of Architecture of the

University of Alicante, where she is currently a lecturer and researcher in the Unit of Urban Design and Regional Planning in the Building Sciences and Urbanism Department. She develops research on urban planning, smart cities and inclusive cities. Her main area of interest is the humanisation of smart cities, that is, the development of models focused on people to improve the design and planning of smart cities. She has published her research results in international conference proceedings and she has also published some papers in international journals. She is also a member of the Research Group on Urban Design and Regional Planning in the Coastal Areas and of the University Institute of the Water and the Environmental Sciences of the University of Alicante.

Carlos Arturo Puente Burgos is a Civil Engineer with a Ph.D. in City and Territory and Master's in Sanitation and Environmental Development. He was Associate Professor of the Pontificia Universidad Javeriana of Bogotá, Colombia and researcher of the Center for Development Projects-CENDEX, currently integrated in the Institute of Public Health of the same University. Carlos has more than 30 years of experience in the design, implementation and evaluation of environmental health projects and is an expert in the development of interdisciplinary studies and community participation.

Albert Rego (alremae@teleco.upv.es) was born in Valencia, Spain on December 13, 1991. He received Bachelor's degrees in Computer Science and Telecommunications Technology Engineering in 2015. In 2016, he received a Master's degree in Telecommunications from the Polytechnic University of Valencia. He is currently working on his Ph.D. under the FPU national scholarship. His research is focused on software defined networks. He is author of several papers and has cooperated in some international conferences, both by reviewing papers and being a part of a committee.

Joan Sapena Femenía (juasafe@doctor.upv.es) was born in Gandía Valencia (Spain) on June 11, 1974. His professional trajectory and training is multidisciplinary. He trained at the Hebrew University of Jerusalem, Universitat de Barcelona, Universidad de Granada, Universitat Politécnica de València and Universidad de Valencia. As a social scientist he is stepping into the arts, sciences and the social level. He trained at the Universitat de Barcelona (2001) in public spaces and urban regeneration, environmental psychology and social stratification and he is working on his doctoral thesis in the field of social robotics. Currently he is developing IoT projects with different technological intentionality and above all creative technology projects.

Sandra Sendra (ssendra@ugr.es) has a Ph.D. in Electronic Engineering and is an Assistant Professor at the University of Granada. She has more than 100 scientific papers in international conferences, journals and books. She is Editor-in-Chief of *WSEAS Transactions on Communications*, and Guest Editor of Special Issues and Associate Editor for several international journals. She has been involved in more than 140 committees of international conferences through 2017 and has participated in 16 research projects.

Giacomo Sorino is an architect with a degree in Building Engineering and Architecture from the Polytechnic University of Bari. He currently works as an architect in the GRAS–Reynés Architecture Studio in Palma de Mallorca and collaborates with the AEDIFICATIO Research Group of the University of Alicante. He is a BIM Specialist/BIM Architect due to having attained the A1, A2, A3, B1 BIM level courses given by EBIME, recognised by Building Smart Spain. During his university career he carried out two study experiences abroad: the first through the Erasmus study programme at the ETSAG University of Granada; the second aimed at the realisation of the master thesis research in BIM methodology and architectural relief in support of traditional architecture in collaboration with the University of Alicante. He took part in the international research work of the AEDIFICATIO Research Group concerning emergency interventions in the Medina of Tetuan and in the thirteenth congress of the CICOP rehabilitation of built heritage.

Daniel M. Taboada Espiniella is a renowned architect, professor and researcher in the Conservation and Rehabilitation field of Historical Heritage. He is a pioneer and a permanent member of the Cuban Vernacular Architecture rescue and has operated since 2002 as the city of Havana's Principal Director of the Gonzalo de Cárdenas Chair in the "Oficina del Historiador".

He has also worked in the Cultural Patrimonial Directorate and in the National Center for Conservation, Restoration and Museology from 1980 to 1997 ("Centro Nacional de Conservación, Restauración y Museología", CEN-CREM), of which he was founder. Daniel has been a Professor Emeritus of the UNESCO Chair for the Integral Heritage Conservation since 2008. He is also the honorary president of the International Council of Monuments and Sites (ICOMOS-Cuba) as well as a member of the International Center for the Conservation of Heritage ("Centro Internacional para la Conservación del Patrimonio", CICOP), of the National Union of Architects and Construction Engineers of Cuba ("Unión Nacional de Arquitectos e Ingenieros de la Construcción de Cuba", UNAICC) and of the Commission for the Career Management and Preservation of the Historical-Cultural Heritage of the San Gerónimo University College of Havana. Out of the many distinctions and awards received throughout his professional career, the following stand out:

- Distinction for the National Culture by the Ministry of Culture in 1983.
- Fernando Ortiz diploma and medal for his contribution to the development of Science and Technology in Cuba by the Academy of Sciences of Cuba in 1987.
- Sauto Theater Distinction, Government Province of Matanzas in 1993.
- City of Havana Award ("Premio Ciudad de la Habana"), Provincial Government and UNAICC in 1994.
- Commemorative medal of the 300th anniversary of the foundation of Regla in 1994.
- Medal of honor and merit from the "Universidad Jorge Tadeo Lozano" of

Cartagena de Indias in 1994.
* Annual research prize by the Ministry of Culture in 1995.
* Diploma for the UNAICC National Architecture Prize in 1998.
* First Prize diploma for the Pan American Federation of Architects Associations in 1998.
* Diploma for the Rehabilitation and Remodeling Award of the Teatro Martí Project in 2002.
* Diploma for the first mention in the restoration category in 2003.
* International CICOP Prize for the Conservation of Heritage in 2005.
* "Provincial Vida y Obra de Arquitectura" Award, UNAICC in 2005.
* Annual award for the most outstanding doctorate in science thesis on January 30, 2008.
* Professor Emeritus of the UNESCO Chair in 2008.
* Honorary researcher of the AEDIFICATIO Research Group of the University of Alicante in 2009.
* University of Havana award in 2013.
* National Prize of Cultural Heritage for the "Obra de la Vida" in 2015.

Francesco Carlo Toso is a practising architect based in Berlin, focusing on building conservation and adaptive reuse. He received his doctorate in the Preservation of Architectural Heritage Programme at Politecnico di Milano. His research and teaching interests focus on analysis and conservation practices for industrial and rural built heritage and landscape. He previously researched landscape issues dealing with the heritage of hydroelectic developments for A2A Energy Group, and was a collaborator of the PARiD landscape research group at Politecnico di Milano on the topic of the reuse of abandoned rural buildings, in the context of the creation of the cultural-agricultural district of Milan.

Pablo E. Vengoechea, R.A., is a registered architect and planner with over 30 years experience in land use and community planning, architecture, education and historic preservation. He is founding principal of v + b Architects and Planners, a multi-disciplinary practice offering services to civic groups, government agencies, the development sector and individual clients. Pablo served as Vice Chair of the NYC Landmarks Preservation Commission for 10 years and was responsible for designating and safeguarding NYC's landmarks and districts. During his tenure, the Commission promoted excellence in design, established preservation policy and designated over 450 buildings and 37 historic districts. He is Visiting Associate Professor of Urban Planning at Hunter College, where he teaches preservation planning, urban design, and planning studios and workshops. He is a member of AEDIFICATIO, a research group based at the University of Alicante, Spain, and the founder of the International Center for the Conservation of Patrimony (CICOP US), a heritage preservation organisation.

Acknowledgements

We would like to thank the citizens and authorities of L'Alfàs del Pi (Alicante) for the collaboration provided during the development of the workshop.

A special thanks to Valentina Prato and Paola Travaglio for the preparation and revision of the book that readers have in their hands.

Preface

Health, wellness and sustainability, from the Mediterranean to the Caribbean

The ambitious title of this book, that I intend to prologue with another transliterated title, brings to my mind the ingenuous thought that the "Mediterranean city" is an economic concession fashioned for marketing purposes. I imagine the Mediterranean city as coastal or near coastal, notwithstanding its name, which indicates that it's surrounded by earth, not water, hence the circumstance of the name of the Mediterranean Sea.

I cannot stop musing, not even for a moment, that of the three characteristics, health and wellness would be physically and spiritually guaranteed due to the existence of the sea more or less nearby. Along with the sea we include the Mediterranean diet, products of the fishing industry that, as has been demonstrated, have positive influences on health and wellness, which as it is a question of mood: if a person believes in it then it is as if it is experienced in reality. Wellness, I believe, is in the mind, in the conditions that people develop, that environment composed of family, neighbourhood, city and Mediterranean city.

I was born in the ultramarine town of Regla, between the coves of Marimelena and Guasabacoa, in the centre of the Bay of La Habana, a city always connected with the sea. My most remote memories are of the sea: the smell of the sea, the sound of the sea. Regla (*Guaicanamar* in the indigenous language, meaning in front of the sea) was a fishing village on land owned by the founder of the village of San Cristóbal and principal neighbour of La Habana in the 16th century, Anton Recio. A settlement that towards the end of the 19th century and beginning of the 20th had evolved by incorporating Spanish immigrants and descendants of forced migration, former African slaves and indentured Chinese workers hired to replace the legally forbidden slave labour. While the Royal Arsenal de La Habana, heir to the one in Cartagena, had declined by then, in Regla it was possible to recreate its activity on a lesser scale, among the smell of tar and the rhythmic sounds of shipwrights working.

If we now have the person capable of taking on health and wellness, that custodian, s/he would be the privileged sustainer of the cited sustainability. To be, one has to be healthy, and along the coast life is better, there is wellness; if those two qualities converge, those people will make any life project that they start sustainable. Sustainability will not emerge from the sea like Venus nor will it fall from heaven like an asteroid, freely, for our enjoyment or annihilation. To be or not, as someone wrote, the game of being or not being is on with climate change. We wager on continuing to exist thanks to sustainability.

The Mediterranean and the Caribbean are in a friendly match – we are islands, island nations, continental coasts around a sea, and you are nations, also around the sea. Inhabited islands, coasts and nations, inhabited cities, what matters is the person that inhabits the territory and the space where one's habitual life is developed, that being is the legitimate heir of a rich, inalienable cultural patrimony. The Mediterranean city and the Caribbean city are possessors of tangible and intangible patrimony of singular characteristics enhanced by the natural environment: scenic landscape, historic patrimony and artistic heritage among others. An irreplaceable tool needed for the development of sustainable territories and cities.

To enjoy a sunset or a moonlit sky, to take pleasure in the Cantigas de Santa Maria from the time of Alphonse X the Wise or a guaguancó solo by Tata Güines, in a paté de foie gras or a fried pork pastry, and a cup of distilled spirits or a shot of rum. We moved from the close of day and nightfall to gastronomy, from the astronomical phenomenon to culinary specialty, the intangible patrimony of the Mediterranean and the Caribbean on both sides of the Atlantic Ocean. And by the side of the Costa del Sol that not for nothing bears a name that is nothing if not ambitious.

If the Mediterranean Sea exists because it's between great extensions of land and the Mediterranean city is in some measure surrounded or partially surrounded by the sea, we point out that the Caribbean Sea exists because belligerent Caribbean groups inhabit it, which is true. The Caribbean city is surrounded or partially surrounded by the Caribbean Sea. And in both cases, often, individuals enjoy the health and wellness that their governments permit. But where the title of the book and the prologue overwhelm me is the third quality, sustainability. That which has to be achieved or succeed should be defensible with virtual tools, including BIM, with non-virtual weapons, with reasons that the heart gives that reason does not comprehend, with whatever.

That the city is sustainable depends on other reasons or tools. In any case, this book, I believe will substantially help to find them. Maybe studying and comparing Mediterranean cities and Caribbean cities, finding similarities and divergences. . . . Perhaps "if starting with citizen participation and the design of such activity as an undeniable element in the construction of democratic cities, active and critical", as I read somewhere, we have in our possession the tools that are missing, now the title of the book would be credible, desirable, necessary and achievable. Let's work on sustainability, the sustainers.

The trio, the three qualities of the title remind me of the indelible impression that left me the surprising contemplation, live for the first time, of "The Spring" by Sandro Botticelli in the Uffizi in Florence. Three beautiful women, three graces, three theological virtues, with arms intertwined in muted silence before the observer. My paternal grandfather Evaristo worked in Cuba at the beginning of the 20th century, as a stonemason on the north façade of the Colón cemetery in La Habana, that ends with three sculptures, Faith, Hope and Charity. In 1928 my father arrived as a young immigrant and settled in Regla. I hope these words that I deliver with pleasure, after a necessary meditation, transmit my gratitude to the authors. I am no longer the same and continue en route. Thank you.

Daniel M. Taboada Espiniella
La Habana, Oficina del Historiador

Prologue

These days, while reading the writings that make up this timely and well-presented work, I have recalled the years in which I walked along many of the marvelous corners of the Mediterranean world, west to east (*ex oriente lux*) and South to North, by land and by sea, and "discovered" the historical realities of its cultures, of its civilizations and societies; and also discovered something essential, something that was a guiding thread for me, for my intellectual and religious vocation, and esthetic education. Some of my most eminent teachers, Luis Diez del Corral, Johan Galtung, Juan J. Linz and Amado de Miguel, were great scholars and lovers of things, cities and the great cultural, political and religious constructs of the Mediterranean and Spain. Others less directly, through lectures and personal encounters, such as Jaime Vicens Vives, Julián Marías, Antonio Tovar, Fernand Braudel, Francisco Rodríguez Adrados and José Maria Blázquez, *ad exemplum*, revealed to me through their work the cultural keys that were treasured, like the study of Hellenic and ancient civilization, starting with a journey (the extraordinary cruise on the Mediterranean in 1933 in which a few of them had the good fortune to navigate).

So as not to turn this note into a forced treatise on the Mediterranean, permit me to recommend five definitive works and a personal one: Ernle Bradford, *Mediterranean. Portrait of a Sea* (1971); Fernand Braudel, *La Méditerranée. Les hommes et l'héritage* (1987); Ramón Margalef (Dir.), *El* Mediterráneo *occidental* (1989); José María Blázquez, *El Mediterráneo* (2006); and the work I consider the most conclusive: David Abulafia, *The Great Sea. A Human History of the Mediterranean* (2011). Years ago I redacted a brief treatise for Sociology of the Culture students, *Cultura y civilización* (2006) that includes a long final essay on the Mediterranean, "El mediterráneo *sub specie temporis*" in which a personal vision and narrative are offered, together with facts, maps and charts on the Mediterranean as a river of cultures that goes back to the origins of *homo sapiens,* which I consider *homo culturalis*, and nearly 30 civilizations. The extraordinary combination of Israel and Phoenicia, Greece, Rome, Christendom, Europe, the Renaissance, the Enlightenment and America, form western civilization, creator, in good measure of the current world system. I refer the reader to the richness in each of these writings if they want to have a complete vision of the research this work offers.

The topic that occupies us and which forms the title of this work, *Health, Wellbeing and Sustainability in the Mediterranean City: Interdisciplinary Perspectives,* is extraordinarily opportune in each of the elements that comprise it. The key is the Mediterranean city. Don't forget that the most ancient walled city of humanity, Jericho, the origin of civilizations, dated by the extraordinary discoverer the archaeologist Kathleen Kenyon, as a small urban unit with several strata, the oldest dating from 11,000 years ago, is in the historic base of this city, the object of this book. The key is in effect, the city, especially, that gentle, extraordinary hill, between Altea and the bluffs of Serra Gelada that is l'Alfàs del Pi. Johan Galtung one of its most prominent sons, teacher of many social scientists and professor during many years in the Department of Sociology of our University of Alicante, "Hellenized" l'Alfàs del Pi in a work program, a sort of intellectual community he denominated Alpha, Delta, Pi (A Δ Π). Ingenious, don't you think? I believe the work presented here, for which Galtung has been interviewed and has examined in other instances, other limited units but otherwise with an enormously rich set of health and welfare variables and facets, is a model *case study*.

I do not want to wear out individuals that will be guided by this well-executed collaboration regarding crucial aspects of our Mediterranean life style as part of the general system of the world city. My mission in this brief note is to celebrate the work and its authors, and invite them to continue this timely research through the sea-lanes opened by this excellent work.

Benjamín Oltra y Martín de los Santos

1   Research on Mediterranean cities

A short introduction of the interdisciplinary perspectives

Antonio Jiménez-Delgado

The Mediterranean city and culture

My city is pierced by the Mediterranean Sea. The smell of the sea anoints the stones, the lattices, the tablecloths, the books, the hands, and the hair. And the sea sky and the sea sun glorify the roofs and the towers, the garden walls and the trees. Where you cannot see the sea you can guess it in the victory of light and in the air that rustles like a precious cloth. In my city from the time we are born, our eyes are filled with the blue from the waters. That blue belongs to us as a portion of our inheritance.

(Miró, 1941, p. 214)[1]

In 1921 the poet Gabriel Miró describes his city, Alicante, and strongly evokes the centrality and relevance of the Mediterranean Sea in the formation of the city and its citizens. The Mediterranean Sea is described as "piercing" the city; it does not only touch it. Where the sea cannot be seen, it is still nevertheless perceived.

The Mediterranean is "a plural area", as defined by Zouain (2010, p. 186) or, according to the well-known description by Braudel (1985), "a thousand things together. It is not a landscape, but countless landscapes. It is not a sea, but a sequence of seas. It is not a civilisation, but a number of civilisations, piled one above the other" (pp. 7–8).[2] Braudel (1985, p. 8) also evokes the image of a "Great Mediterranean", a sea without precise boundaries.

The Mediterranean can be considered time and again a complex crossroads of people, cultures and events (Cancila, 2008). This sea can come across as a "frontier" which unites and at the same time can feel "as much as a divide" (Cassano, 1996, p. 22). It is therefore possible to talk about a "Mediterranean culture" that intimately binds all cultures, people and events. Holding the sense of unity and complexity together, this culture, rooted in the Mediterranean territory and in its people, provides them with a sense of identity. As Zouain (2010) wrote,

the continuous trace that the Mediterranean has left on people is constituted by a deep overlapping of cultural layers, behaviours, philosophies and shared religions that, although they have often opposed each other, have always ended up rediscovering and mingling in that common magma formed by the Mediterranean Sea and the lands included in the limits of its geographical space.

(p. 186)

The Mediterranean city is part of this culture. Inevitably and strictly linked to the sea, Mediterranean cities are the result of centuries of history that can be read through the cultural narratives built by their inhabitants and stratified over time. The Mediterranean culture is the common denominator of a wide space of local and global interrelationships where the city becomes the meeting point historically approached by the sea.

In recent decades there have been numerous studies on Mediterranean cities and their development from an urban, social and economic point of view (for instance, Leontidou, 1990; Cardarelli, 1987, 1990; Pace, 1996). However, is it possible to talk about "the Mediterranean city"? What elements characterise it? Is it ascribable to a single model? The answer seems at the same time both direct and complex. Following the reflection by Pace (2002), despite the presence of a great variety of heterogeneous Mediterranean urban realities and overcoming the "Mediterranean Myth" based on a mythological, romantic or vernacular heritage, it seems possible to propose the existence of a Mediterranean city on the basis of common physical, morphological and architectural elements, but also social and cultural.

Mediterranean cities often underwent accelerated urban development, frequently due to complex migratory movements, with the consequent creation of tensions in the territory and the application of unsustainable development models. This book was born from the belief that the Mediterranean city is the bearer of an extraordinary wealth, the result of a historical cultural mix and permanent territorial tensions – highly relevant issues nowadays. These tensions can be enriching, but require strategies to place human beings and their health, in the broad sense, at the centre of interest. In order to analyse the Mediterranean city as a healthy and sustainable space, an interdisciplinary perspective is necessary to help us define those elements that, within an incontrovertible complexity, allow us an accurate approach to the aims that the United Nations establish for sustainable development.

The AEDIFICATIO International Working Group

The present volume is the result of a close collaboration between researchers in different fields and countries belonging to the AEDIFICATIO International Working Group. In 1996, with the aim of coordinating research on the management and preservation of architectural heritage, the author of this short introduction, supported by the University of Alicante, convened an informal international working group, later registered in 2008 as the AEDIFICATIO International Working Group for Construction, Technology, Research and Development, coordinated from Milan, New York, Granada and Alicante.

The group is the result of an interdisciplinary approach and consists of around 30 researchers working in various fields ranging from architecture and preservation of the architectural heritage, engineering, urbanism and city planning, geography, sociology, history and art history. While some of the group's members are self-employed, others are staff members of the following institutions: Universidad de Alicante, Universitat Politècnica of Valencia, Universidad de Granada, Universidad Complutense of Madrid, Politecnico of Milan, The City University of New York–Hunter College, Pontificia Universidad Javeriana of Bogotà, Oficina del Historiador de la Ciudad de La Habana (Cuba) and the Italian Ministry of Cultural Heritage (Museo Storico e Parco del Castello di Miramare, Trieste).

About the contents of this book

The 12 essays comprising this volume present the most recent and interdisciplinary research on the Mediterranean city, developed starting from the case study of L'Alfàs del Pi (Spain). L'Alfàs del Pi represents a municipality of great geographical uniqueness, with a population made of a mixture of local inhabitants and Northern European residents attracted by good weather conditions and sea.

Despite the complexity of the wider Mediterranean context, the book is the result of varying approaches to the subject and looks at diverse aspect of the Mediterranean city: the sociological analysis of its territory; the analysis of its environmental and cultural heritage elements; the examination of its urban design and landscape character; the assessment of its public health and sustainable mobility; the use of new technologies as development tools; and citizen participation.

The general aim of this research is to design a sustainable, healthy and innovative city model, which could offer high standards of living and a global reference in the Mediterranean area. As mentioned at the outset, the main characteristics of the research are interdisciplinarity, which allows an integral analysis of the territory and an international approach, thanks to the participation of universities and researchers from different cities and countries.

The book is divided into two parts. The first, "Context of the Research: A Theoretical-Methodological Approach", of a general nature, tries to establish the theoretical and methodological tools to analyse a specific territory of the Mediterranean coast, including social, cultural and economic structures, as well as heritage elements of both a tangible and intangible nature. The methodological strategy is always made starting from an interdisciplinary approach and considering the framework of health, wellbeing and sustainability in the development of cities. In the second part, "Technical Reports and Proposals of Intervention in the Territory", an analysis of specific aspects is exposed, often using a Mediterranean city – L'Alfàs del Pi (Spain) – as a concrete case study, and presenting various methodological tools.

In Part 1, Pablo de-Gracia-Soriano et al. discuss the relevance of the analysis of the social, cultural and economic structures as a basis to understand the territory where the activity of the city takes place. On the other hand, a paper by Carlos Arturo Puente Burgos et al. examines interdisciplinarity as a tool of irrevocable importance in this research work, focused on the territory as a complex space of coexistence with kaleidoscopic and changing elements in a local and global field.

The study for the preservation and promotion of the tangible and intangible heritage brings us closer to those elements that make up the identity of a territory. As discussed by Carlo Manfredi et al., the heritage – intended in a broad sense – helps us understand and contextualise the life of people who inhabit the Mediterranean city. "The practice of the sociological imagination requires an awareness of these popular stories of the personal issues of lived experience, and the construction of connections with the account of the epoch" (Bauman, 2014, p. 5). Identity is a social construction that structures our way of life. Identity and memory are necessary concepts in understanding our Mediterranean DNA.

In this first part of this book, the concepts of Smart City and healthy destinations are also explored. A paper by Raquel Pérez-del Hoyo focuses on the use of new technologies for a sustainable contemporary development of the Mediterranean city, while Carlos Arturo Puente Burgos et al. define the idea of "healthy destination" in the global and particular context of the Mediterranean Sea.

In Part 2 Diana Jareño-Ruiz et al. examine the citizen participation as a strategic axis of the work methodology, which is useful to build democratic and active cities and understand the needs and perceptions of people who inhabit the territory. Through the analysis of mental maps, Pablo E. Vengoechea explores the environmental, cultural, patrimonial etc. elements that are uniquely perceived by the inhabitants of the territory, thus establishing strategies for the subsequent intervention and enhancement, such as the future transformation of Route N-332 into the Great Boulevard of L'Alfàs del Pi.

A paper by Armando Ortuño Padilla and Jairo Casares Blanco focuses on the urban and interurban mobility in the municipal area of mid-size Mediterranean cities with the aim of achieving a healthy and sustainable city. Strategies associated with green corridors and quality spaces in urban and interurban mobility are established. The agricultural and historical landscape as a cultural element is analysed by Francesco Carlo Toso as part of the identity of the territory and in relation to the inhabitants and their ancestral roots connected to the Mediterranean area.

Laura García et al. present a direct application in the field of wellbeing and health discussing the monitoring of domestic environment useful to optimise the resources available within the private sphere, while Giacomo Sorino et al. describe methodologies for the preservation of the architectural heritage. In particular, the BIM methodology generates an efficient documentation for an adequate constructive intervention in building something of historical value. In the final paper, Lorena Parra et al. present the use of new technologies for the management of water in urban environments, in order to optimise this basic and scarce resource in most cities of the Mediterranean coast.

The Mediterranean city represents a space that must be preserved and constitutes, at the same time, a fundamental educational tool. In fact, the ancestral network of routes across the Mediterranean Sea – "from Algeciras to Istanbul", as Joan Manuel Serrat sings (see Figures 1.1. and 1.2) – and endless connections between cities make up an extraordinary learning space for new generations. Connected by water, the traveller saw the city from the "waterfront", which identified without any doubt the place where he came. A place full of content: the Mediterranean city.

The Mediterranean city can be used as a useful educational resource. The discovery of the heritage of an initially unknown city is capable to generate in

Figure 1.1 Hagia Sophia, Istanbul.

Figure 1.2 View of the Strait of Gibraltar from Gaucín, Serranía de Ronda, Andalusia.

citizens attitudes of tolerance, respect, admiration and solidarity, which are indispensable values in achieving true solidarity and cooperative education.

The Report to UNESCO International Commission on Education for the Twenty-First Century (1996, *Learning: The Treasure Within*) indicates that the four pillars on which the education of the 21st century should be based are: learn to know, learn to do, learn to live, and learn to be. "Knowledge is the food of the soul", said Plato, and "We learn not in school, but in life", Seneca would say. We find therefore a privileged *learning space* in the Mediterranean and its cities, its people and habits. The health of the body and soul of each individual is the centre of the present volume; the broad stage is the Mediterranean Sea and more concretely the city is the space where life should become sustainable.

Notes

1 Translated by the author of the introduction from the original text: "Mi ciudad está traspasada de Mediterráneo. El olor del mar unge las piedras, las celosías, los manteles, los libros, las manos, los cabellos. Y el cielo de mar, y el sol de mar, glorifican las azoteas, y las torres, las tapias y los árboles. Donde no se ve el mar se le adivina en la victoria de la luz y en el aire que cruje como un paño precioso. En mi ciudad, desde que nacemos, se nos llenan los ojos de azul de las aguas. Ese azul nos pertenece como una porción de nuestro heredamiento [. . .]".

2 Translated by the author of the introduction from the original text: "Mille choses à la fois. Non pas un paysage, mais d'innombrables paysages. Non pas une mer, mais une succession de mers. Non pas une civilisation, mais plusieurs civilisations superposées".

References

Bauman, Z. (2014). *What use is sociology? Conversations with Michael Hviid Jacobsen and Keith Tester*. Cambridge: Polity Press.

Braudel, F. (1985). *La Méditerranée. L'espace et l'histoire*. Paris: Flammarion.

Cancila, R. (2008). Il Mediterraneo: storia di una complessità. *Mediterranea Ricerche storiche*, *13*, 243–254.

Cardarelli, U. (Ed.). (1987). *La Città Mediterranea. Primo rapporto di ricerca*. Napoli: Istituto di Pianificazione e Gestione del Territorio.

Cardarelli, U. (Ed.). (1990). *La Città Mediterranea. Secondo rapporto di ricerca. Aree urbane e sistemi metropolitani nel Mezzogiorno d'Europa*. Napoli: Istituto di Pianificazione e Gestione del Territorio, Dipartimento di Pianificazione e Scienza del Territorio.

Cassano, F. (1996). *Il pensiero meridiano*. Bari-Roma: Laterza.

Leontidou, L. (1990). *The Mediterranean city in transition: Social change and urban development*. Cambridge: Cambridge University Press.

Miró, G. (1941). *Obras Completas de Gabriel Miró. Vol. 8. El humo dormido; El angel, el molino, el caracol del faro*. Barcelona: Tipografía Altés.

Pace, G. (Ed.). (1996). *Sviluppo economico e urbano delle città mediterranee*. Napoli: Istituto Italiano per gli Studi Filosofici.

Pace, G. (2002). *Ways of thinking and looking at the Mediterranean city*. Napoli: Istituto di Studi sulle Società del Mediterraneo.

Zouain, G. S. (2010). ¿Constituye el patrimonio cultural inmaterial un lenguaje común para el Mediterráneo? *Quaderns de la Mediterránia*, *13*, 185–188.

2 Analysis and definition of the social, cultural and economic structure and dynamics

*Pablo de-Gracia-Soriano, Diana Jareño-Ruiz,
María Jiménez-Delgado*

Considerations for a sociodemographic diagnosis

Carrying out a sociodemographic diagnosis of a municipality is a necessary action when research or policies are to be developed. These policies are significant because they directly or indirectly modify the population structure and dynamics. Experience tells us that city councils or entities that work in a concrete municipality should not dispense with this effort. The main reason for this sociodemographic diagnosis is to have a greater degree of knowledge of social reality in order to face the possible risks of any political or scientific action.

One of the most relevant shortcomings that is frequently observed in the development of municipal diagnoses is the low availability of basic population data at a sociological level. This is often a problem when trying to make a diagnosis. The lack of data for diagnosis requires actions that contribute to having data and information that can make the diagnosis viable.

This scenario suggests taking into account the need for at least two types of strategies. The first of these strategies is to promote the opportunity that the city councils have to obtain official data from state entities. In the case presented in these pages, for example, the National Institute of Statistics, the Valencian Institute of Statistics or the Diputación de Alicante can respond to this lack through their data banks. In this way, the city councils would have available official data concerning the municipality. In addition, once the data is obtained and structured, city councils would have the possibility of making the data public (accessible and manageable) through a statistical section on the official website. This last management facilitates both the citizens' degree of knowledge about their environment as well as the opportunity to access official data for any research that may be carried out in the municipalities.

Through these organisations, and the means available to the local government at present, it is possible to have basic information about the population, such as natural population movements (births and deaths), residential variations, income levels, the state of health and physical and mental pathologies or diseases, the uses of time, the risk of poverty and social exclusion, the use of information and communication technologies and displacements. All this information must also allow crossing with basic sociodemographic variables and other factors of recognised relevance, such as sex, age, marital status, level of education, place of origin, nationality, profession, occupation, employment status, place of residence, etc.

On the other hand, these indicators can be estimated significantly or in a relevant way in another manner: the data could be produced from the municipality itself. In this sense, it is common to find municipalities that produce data through different studies, quantitative and qualitative, which offer an opportunity to increase knowledge and the probabilities for municipal improvement. Frequent cases, although perhaps not too many in the Spanish cases, are surveys and barometers of public opinion as well as qualitative research that allows knowing the local government, from possible sociopolitical conflicts and their processes to feelings or discourses generated by certain populations or groups of interest. All this data contributes to having a better and more profound image of those who relate to the municipality.

In summary, not all municipalities have detailed information that is open and accessible. In fact, the most common reality is to find data scattered among different sources that are more or less official in nature. There are also cases in which local governments have the information, but in the application process they are reluctant to share it, sometimes due to legal issues, and sometimes for political reasons.

According to the above, sociodemographic diagnoses become a first-order tool to plan and manage municipal actions, as well as to know the contexts in which actions and decision-making processes will take place. For all the above, these sociodemographic diagnoses must be understood from a mixed perspective, both quantitative and qualitative, which allows us to diagnose more deeply the current situation of any municipality regarding certain issues, as well as its evolution in certain time periods.

Case study: quantitative sociodemographic diagnosis of L'Alfàs del Pi

Structure and dynamics of population

Distribution of the population in the municipal space and in time. The population of L'Alfàs del Pi has doubled in the last 20 years. On January 1, 2017, according to the municipal census offered by the National Institute of Statistics, the municipality had 21,494 residents (INE, 2017b). Although there are more women than men (68 more), the difference allows us to speak of a population that, overall, is balanced by sex.

Currently, we observe a demographic density of 1,116 inhabitants per square kilometer, which represents a high population density. This figure places L'Alfàs del Pi as one of the most densely populated municipalities in Alicante (ninth position) and Spain (position 252 in a ranking of population density for all 141 municipalities in the province of Alicante, and 8,119 in Spain).

In general, the migratory balance has been positive with the exceptions of years 2001, 2008, 2012 and 2013, in which it has been practically zero (for the first three) and negative (in 2013). This data show us that especially during the first years of the 21st century, the immigrant population contributed to population growth. L'Alfàs del Pi is in general considered a receiving municipality rather than a transmitter. The exception is found in 2013, when the migratory balance was −329, so there was a slight decrease in the absolute number of population motivated by the departure of more people than those who entered the municipality and only cushioned by the demographic events of births and mortalities. Despite this exceptional fact, one can speak of a permanent population increase in the last decades, especially in the first decade of the 21st century, in which migratory processes have played an important role. As we can see from the year 2004 record, population increase was not caused by the migratory process alone. In 2004, in fact, despite migratory balance reaching its maximum value, the population was reduced by almost 2,000 people. The loss in population was possibly caused by the mortality being higher that year.

Regarding the spatial distribution of the population, the municipality has 21 territorial nuclei (INE, 2017c), of which three have more than 5,000 residents (L'Alfàs del Pi (center), L'Albir and Pla Partial Platja), while the other 18 have smaller populations. Considering the geographical division proposed in the project (see Figure 2.1), the first corresponds to the census sections number 4 and 5, with more than 20% of the population, the second includes sections 1, 3, 7 and 8 with almost 40% of the population concentration and the third zone, which includes sections 2 and 6, has a total of 8,637 residents.

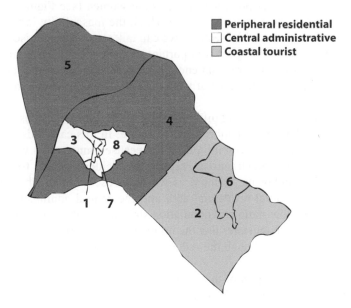

Figure 2.1 Municipal areas.

Source: Own production.

Population pyramid

The analysis of the population pyramid shows that L'Alfàs del Pi has a population structure that, although certainly balanced in terms of the sex ratios, presents a generationally *mature* structure. This is because the bulk of the population (73.4%) is over 29 years old. Given its bulbous shape, we can say that it is a regressive population since the base is smaller than the rest. This fact can be interpreted as a low presence of infant-juvenile population which is generally associated with a low birth rate. Given the tendency of the population to age in a certain way in developed regions, the *narrowness* of the base of the pyramid shows us future risks for the adult population. These risks emerge especially in regard to the dimension of care (Durán, 2018), which as in most populations of this kind, is becoming an increasing necessity. If the population pyramids are analysed from the census sections and the geographical areas previously mentioned, some evidence can be observed that may be interesting when taking into account any type of policy in the municipality.

By census sections (see Figure 2.2), the 4th and 6th are those that have a regressive population structure, which reveals regions inhabited by an aging population. On the contrary, sections 1 and 8, although they maintain a structure in the form of a bulb, do not reach the inverted pyramid forms. Sections 1 and 8 are therefore regions with a lower average age, which makes the population more mature than aged.

Another relevant issue is that in all the sections, except for sections 3, 4 and 6, there is a greater presence of men than women (see Figure 2.3), with census sections 1 and 5 standing out, in which the masculinity index is 1.03. This means that in most of the sections we can talk about masculinised regions, especially the latter, although it is important to make a prudent reading of this data since there is not a significant enough difference in numbers. Section 3 is where there is a greater presence of women in comparison, with a masculinity index of 0.93.

When taking a first glance at different geographic zones, the coastal tourism zone is the one that shows a more aged structure. In this case, this region has the shape of an inverted pyramid, possibly because immigration was initially motivated by sun and beach tourism, especially in elderly people from outside the municipality and/or the country, as we will observe later. The other two zones are the peripheral residential area and the central administrative area. Both these areas show stationary population structures (the pyramid has a more tubular shape), which indicates that the infant-juvenile population is in balance with the rest of ages, although the reduction of the base of all the pyramids is visible.

A final noteworthy aspect of this data is that in the three areas there is a widening around the ages of 35–55 years, which may be explained by the migratory process of individuals migrating because they need labour. The lack of data on first-order demographic phenomena for the municipality (births and deaths mainly) means we are unable to explain these population peculiarities.

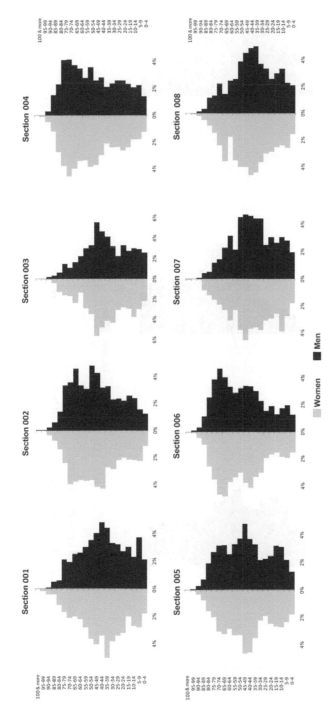

Figure 2.2 Population pyramid by census section.

Source: Own production based on data from INEBase: "Revisión del Padrón municipal a 1 de enero de 2016".

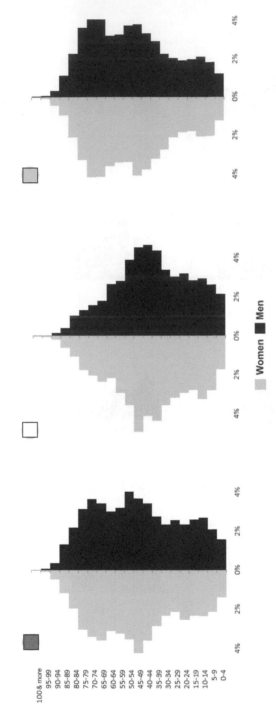

Figure 2.3 Population pyramid by municipal area.

Source: Own production based on data from INEBase: "Revisión del Padrón municipal a 1 de enero de 2016".

Population structure according to place of birth

L'Alfàs del Pi is one of the Spanish municipalities where we observe a higher presence of foreign population that was born in the municipality or in the country (see Figure 2.4). Specifically, 43.5% of the residents were born in Spain, compared to 56.5% who are native to another country. Only 1 in 10 people is native; that is, he or she was born in the municipality. Continuing with our regional analysis, we notice that the most polarised area in terms of the population dichotomy of Spanish-foreign origin is the coastal tourism zone. Meanwhile, the central administrative area is characterised by more Spanish population than foreign individuals.

In the three regions we found a prevalence of population born in another country of the European Union compared to other origins. The more remarkable countries of origin are Norway (13.5%) and United Kingdom (9.7%). Those two nationalities amount to a high percentage of population with respect to the total population in the municipality. In fact, those two countries citizens are in higher proportion in L'Alfàs del Pi than those born in the municipality itself.

As for the distribution of the population in the territory, people born in Spain are concentrated in the central administrative area, while those born abroad reside in the coastal tourism zone. Specifically, people born in a non-EU country are distributed in the municipality in a more homogeneous way, throughout the three zones analysed. However, people born in Africa, Asia and Oceania have more concentrated values, especially due to their low presence in the peripheral residential area; the same applies to those who were born in the municipality, who are concentrated in the central administrative area.

Households and homes

The last census regarding households conducted in Spain was in 2011. This census was "viviendas" and "hogares", indicating two types of different households. According to this last census conducted in Spain, in L'Alfàs del Pi there were 8,530 households, corresponding to 69% of the total number of houses (12,406), while the municipality is made up of 19% secondary housing. The rest (1,418, or 12%) corresponded to empty houses. This last aspect should be pointed out since it would be interesting to know the reason why the houses are empty, as well as the context of property that surrounds them. The vacant homes could present an opportunity for the municipality to satisfy social, economic, cultural and environmental needs, depending on their location and concentration in the territory. Regarding the main dwellings, which correspond to the number of households, we see that two-thirds of these dwellings are property-owned. More than half of these dwelling are in the process of being paid, while the other half is already fully paid; 20% of households (1,740) are for rent.

On the other hand, regarding the composition and size of households, the 2011 census found that just under a third of the 8,530 households are occupied by a single person, 37% of homes are composed of two people and the rest have more than two people. Approximately 38% of households with a family nucleus (6,003)

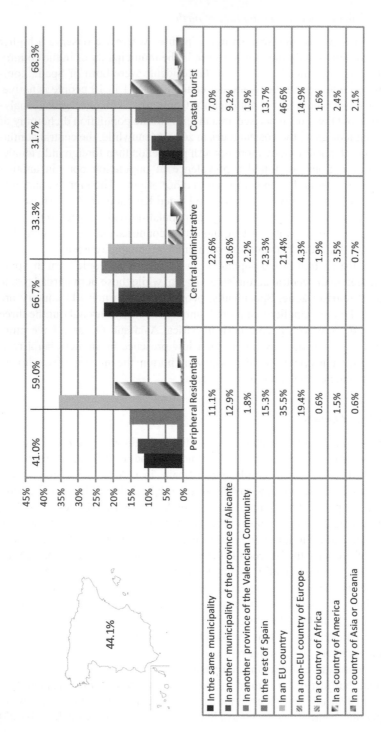

	Peripheral Residential	Central administrative	Coastal tourist
In the same municipality	11.1%	22.6%	7.0%
In another municipality of the province of Alicante	12.9%	18.6%	9.2%
In another province of the Valencian Community	1.8%	2.2%	1.9%
In the rest of Spain	15.3%	23.3%	13.7%
In an EU country	35.5%	21.4%	46.6%
In a non-EU country of Europe	19.4%	4.3%	14.9%
In a country of Africa	0.6%	1.9%	1.6%
In a country of America	1.5%	3.5%	2.4%
In a country of Asia or Oceania	0.6%	0.7%	2.1%

Figure 2.4 Geographical composition according to the birthplace of the resident population.

Source: Own production based on data from INEBase: "Revisión del Padrón municipal a 1 de enero de 2016".

have children in the home, while the rest are homes where couples reside. In addition, in those households with the presence of children (3,273), 77% are formed by a couple, while in the rest of the cases the nucleus is formed by only one of the parents, with part or all of their offspring.

It is thus obtained that there are 1.75 inhabitants per dwelling, or in another way, 0.57 dwellings per inhabitant. However, according to the data we have explained above, we see that this figure is not distributed equally for the population as a whole in the municipality since according to the data, we obtain an average of 2.3 inhabitants per main dwelling. Finally, in line with the comments above, 45% of the residents are married, while 41% are single, 6% are widowed and the rest are separated or divorced. All this data indicates that more than half of the municipality's population is alone or, if it is in a couple, it is not under matrimonial institution.

Education and health

Levels of education and qualification

Age and level of studies maintain an inverse relationship, since we see that the highest grades achieved in terms of educational level are concentrated in the young population, or in other words, the population groups with more advanced ages have lower levels of regulated education. Although there is such a correlation, we cannot draw conclusions since we find in the municipality's data that approximately 20% of those with third-degree studies are persons over 64 years of age, or that around 35% of those who do not have formal studies are less than 64 years old. The majority of residents consulted in L'Alfàs del Pi was 16 years of age or older, had reached second grade with respect to educational level and only one of every ten inhabitants had completed first grade. In the municipality, we found more than 1,000 people who did not have studies.

In general, the main difference that we find in the levels of studies according to sex is that women reach higher levels of education than men, which supposes a professional and labour opportunity for the women of the municipality, who can choose higher positions. Similarly, the data shows the urgency of knowing why men have a lower level of education than women.

Methodological prudence is necessary in this case as we analyse a slightly divergent territorial framework. While looking at this slightly different territorial framework, we can observe a wage difference between both sexes in the region of the Marina Baixa. This fact, to be emulated in the municipality that we are analysing, would reflect a first and clear symptom of inequality due to sex. However, as we indicated, in the absence of specific data for the municipality, we cannot draw conclusions.

In general, this scenario regarding the level of population studies suggests the opportunity for the local government to act in the context of the inequalities generated, as well as the realities found, taking into account the gender issue and the relationship between studies, work and remuneration. Finally, according to the Territorial Data Bank of the Valencian Institute of Statistics (IVE, 2017), it should be noted that the municipality has provided a municipal museum and a library.

Dependence and some demographic indicators

The dependency rate is an index that represents the weight of the population over 64 years of age and under 16 years of age, compared to the population between 16 and 64 years of age. In this way, the rate allows us to know how many people who are in potential dependency status there are for each of the people who fit into a low-dependency model.[1] We see that in the municipality, for the year 2015 we find a dependency rate higher than 60%; that is, for every ten *non-dependent* people, we found six potentially dependent. This rate has been increasing since 2004, coinciding with the increase in the population over 64 years of age in the municipality, as reflected in the aging index. This index shows that for the year 2015, the municipality has almost twice the population over 64 years of age compared to those under 16 years old. In addition, the longevity index (weight of the number of people over 74 years of age with respect to the number of people over 64 years of age) has also grown from 2004 to over 40% in 2015.

As shown in the graph above, the general dependency rate is correlated with that of the population over 64, while the population under 16 has been decreasing since 2004. This fact confirms the progressive aging of the population, which is fueled by the decrease in maternity rates in recent years, producing a scenario where there are fewer and fewer people under 4 years of age with respect to all women of reproductive age (normally between 15 and 49 years old). From this data we see that, as discussed in the first section, demographic inflection between 2003 and 2004 can be found.

Employment and tourism

Economic activities

The last five years has been marked by the presence of an interval between 1,800 and 1,900 companies in the municipality. In 2016, the number of companies was 1,841, forming an economy based mainly on the services sector (83% of the main activities) and construction (around 15%). The industrial sector, on the other hand, comprises just more than 1% of the total companies in the municipality, with 30 companies in this sector. Finally, it should be noted that, according to the Agrarian Census, through the INE Territorial Statistics Report (2017a), L'Alfàs del Pi had 100 farms in 2009.

As for the different activities that are produced and developed in the municipality, trade, transport and hospitality are the groups in which more companies appear, and therefore boost the municipal economy. It would be interesting to know the volume of income and activities of the companies that appear on the scene.

From the data found, we can talk about three different scenarios that have been happening in the alfasina companies over the last five years, according to their microtendence. On the one hand, there are activities that have been decreasing in recent years: this is the case of construction (-17%), industry (-14%), financial and insurance activities (-10%) and trade, transport and hospitality (-9%). On the other hand, there are three activities that have grown during this period: the information and communications sector ($+13\%$), real estate activities ($+27\%$) and other personal services ($+28\%$). The rest of the activities, despite their variations, remain stable throughout the observed period.

Personal services are the activity that is growing the most, and this remains consistent with what was previously commented on the structure and dynamics of the population: as the population ages, care-related activities are growing. This growth in personal service needs requires socio-labour regulation, due to the social characteristics that surround and contextualise these activities, especially from gender perspectives and origin of the working people.

Travel times and characteristics

In the case of L'Alfàs del Pi, there are some statistics and data missing from the municipality. Scientific discourse is also limited, which makes our diagnosis in this chapter useful for the present time but somewhat imprecise. The focus of this section is on the relationship between work and time based on three variables: relationship between the municipality of work and residence, the means of transportation used to go to the workplace and the time allocated to move from home to the workplace.

These variables present an image of the types of displacements that are generated in the municipality from those who reside and are registered in it, obtaining that approximately half of the census population generates internal displacements, while the rest must leave out since your place of work is elsewhere. More than three-quarters of the population takes less than 20 minutes to reach their workplace, which indicates that, in general, even those who do not work in the municipality itself can still live in nearby regions and commute. Finally, an interesting fact is that approximately 60% of the population, despite this proximity between housing and the workplace, makes use of private transport, compared to 12% using public transportation. The reasons can be very diverse, so to draw conclusions in this context requires investigation into the causes, from municipal deficiencies (lines, frequency, prices, etc.) to family and/or personal contexts, to gain a better idea of why one type of transportation was chosen over the other.

Unemployment in the municipality

Knowing unemployment and its index is generally complicated at a municipal level, since, although the number of people seeking employment is usually known, the number of people who make up the active population is less known. This context usually produces a scenario in which an estimate of the number of people unemployed is usually made, taking into account those people who work in the municipality, and whether or not they are counted in it. The use of surveys in this regard is common, although the risks of making use of these data are evident. In this section, references are made to an estimate and are not yet based on real values. In the absence of other indicators, we cannot know if underemployment data is an overrepresented value (there would be many people working in the municipality but counted elsewhere) or underrepresented.

Having said this, unemployment in the L'Alfàs del Pi municipality was accentuated with the progressive arrival of the economic crisis, until doubling in 2008–2009. The crisis has been reflected in new structural reality since then. Since 2012 unemployment has been decreasing at an average rate of 64 people per year, which is around 5% per year. 2017 was one of the greatest drops in this indicator since unemployment has been reduced (at least until October 2016) by 93 people, which is 8.3% less than the previous year (see Figure 2.5).

The sociodemographic profile of people in unemployment shows that there is a clear correlation between age and unemployment in the municipality; this situation worsens as the population ages. This fact is necessary to emphasise because L'Alfàs del Pi is a population that tends to age structurally, so the rate of unemployment may increase in the future. Another element that requires attention is the fact that, except in the case of younger population under 25 years of age, unemployment is concentrated more in women than in men. We can speak of a remarkable feminisation of unemployment in L'Alfàs del Pi because the feminisation index is 1.22; that is, for every eight unemployed men, there are ten women in the same situation. However, we must keep in mind that it is women

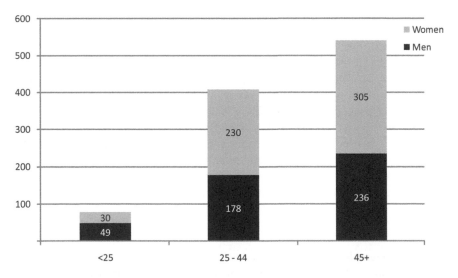

Figure 2.5 Number of unemployed people, by sex and age.

Source: Own production based on data from INE, "Estadísticas Territoriales" (2016).

who have higher educational levels. If we focus on the population over 24 years old, unemployment becomes worse for women, until we reach a feminisation rate of unemployment of 1.29. In other words, we find that in L'Alfàs del Pi, women are between 20% and 30% more likely to be unemployed than men, regardless of the sector in which they work. By sectors, unemployment is distributed in a very similar way to the general structure.

Tourist offerings

According to the latest data available at the time of the consultation (offered in 2014 by the Conselleria d'economia, indústria, turisme i ocupació of the Valencian Community), tourist offerings in L'Alfàs del Pi included 11 hotels, 4 hostels, 1,324 apartments, 1 pension and 207 restaurants. In total, the municipality has the capacity, according to the registers, to host a third of its population, that is, 7,334 tourists. On the other hand, the 207 restaurants in the municipality could serve 10,749 people, just over half of its current population.

The municipality in virtual social networks

The studies referring to tourism are a source of interest, but they are not usually specific to the municipalities, because they cover broad or general regions at the provincial level. One of the main sources available in the production of knowledge in this area for the whole of the Valencian community is the Valencian Institute of Tourist Technologies (INVAT.TUR). Although specific data of L'Alfàs del Pi has not been obtained, INVAT.TUR periodically offers a barometer showing virtual social networks within the framework of touristic presence in the Valencian community. The last barometer consulted[2] corresponds to the first semester of 2016 and offers some data that are relevant to know the notoriety of the municipality in the Valencian group, also paying attention to different digital-virtual platforms (Instagram, Twitter, Facebook and YouTube). Knowing the presence and interest that the municipality has in these networks also means that it is possible to indirectly know the means through which the municipality can launch messages with greater scope.

The Costa Blanca is the tourist brand that has the most fan-followers in digital-virtual social networks, with 59,579 fans on Facebook, 19,332 followers on Twitter, and 11,973 fans on Instagram. However, regarding reproductions of tourist brands' videos on YouTube, those of the City of Valencia stand out (with more than half a million reproductions), followed by those of Alicante (301,014 reproductions), the Costa Blanca (197,079) and Benidorm (192,604).

The VII barometer reveals that the municipality has the greatest interest in Twitter (2,383) and Instagram (1,437), while it does not appear in the other two social networks (depending on the number of followers). Although in general the municipality falls in the last positions as tourist destinations of the Costa Blanca, the absolute figure is an element, in this case, that has to be taken into account as a positive value for the municipality.

Summary

This chapter showed a practical example of a basic sociodemographic diagnosis. The classification of this data has been made from four sets or dimensions of human life in a specific territory. On the one hand, it is necessary to know what are the population structures that exist in the municipality as a whole, and as far as possible in each social or geographical space. This is important because it allows us to know basic characteristics of the population such as sex, age or origin. In addition, it allows us to know how this structure has evolved over the years. Any design of public policies, as well as any possible intervention in a specific region, must be taken in account to obtain data for a plan of action that is coherent with the sociodemographic reality of the place.

Secondly, attention has been paid to the presence of housing in the territory. In this section, a double reading should be considered, since on the one hand they can be interpreted as a configuration of the physical space or the projection of a general plan, and on the other hand they should also be considered from their social component. In this sense, it is important to highlight aspects that refer to the composition of households as well as the distribution in the territory of different socio-economic profiles and social phenomena that occur in the relationship between social and physical space. Examples are spatial segregation, empty housing, urban sprawl or gentrification processes. These dynamics can be observed and analysed from the data of these dimensions, especially when they are supported by a census or by a research representative of the region's general configuration.

The third block that has been presented refers to education and health issues. Aspects such as dependence levels, the most common diseases and the educational levels of the population allow us to observe and analyse the basic needs of the region, both social and physical. This last aspect allows us to draw a link between training and income level and to know habitual or unequal situations from different socio-economic profiles. The analysis of the evolution of the indicators in these dimensions also allows us to evaluate public policies and to know significant contexts that could have a certain predictive character.

Finally, the fourth block refers to the socio-economic aspects of employment, such as tourist offerings. In this sense, a diagnosis must contemplate the structure and dynamics of the active population in the region studied, as well as the economic activities that take place in it. In the case of the Mediterranean regions, although not necessarily in all the municipalities, knowing the tourist offerings helps us analyse both the economic activity as well as the population dynamics and its stationary fluctuation. These aspects sometimes pose serious headaches for local governments, and that is why knowing this data allows us to prevent future problematic scenarios.

In summary, the dimensions that have been presented here are all measurable. The social component stands out more in some than in others, in which the focus is placed on the territory and the physical space in general. A sociodemographic diagnosis, from the sociology presented here, must capture social and demographic phenomena: population, economic, political, organisational, leisure, consumption and tourism. A sociodemographic analysis cannot be understood only from demographic facts or population movements (Carratalá-Puertas, Ruiz-Callado, de-Gracia-Soriano, & Casasús-Chorques, 2018). The social part is an important one and it is the one in which all the possible aspects of the social relationships are gathered. Precisely because of its social nature, a sociodemographic analysis must be considered from multiple perspectives and areas of knowledge.

Certainly, predominantly physical aspects can escape a diagnosis of this type, but it is considered more and more necessary to integrate these facts, phenomena, structures and dynamics since they are also the product of social relations: sometimes political, sometimes economic, sometimes cultural, sometimes social, in a narrow sense. Perhaps they should be called sociological relations in order to distinguish them from the ideological use of the word "social". Examples of these aspects with a strong physical component may include physical heritage, cultural heritage, urban or territorial planning, crop area, forest area, communications, transport, natural parks, etc.

In this sense, a sociodemographic diagnosis that we consider adequate (that is, that has a sufficiently solid base to analyse and act in the sociological realities of the Mediterranean) must consider multiple dimensions and variables (see Figure 2.6). That is why, in addition to disengaging the dimension of human,

Dimensions. territorial capital	Demographic and social	Cultural and services	Natural and economic
Variables and indicators	Demographic indicators	Fire stations	Agricultural operations
	Dependency index	Hospital beds	Blue flags
	Middle Ages	Libraries	Companies according to main activity
	Migrations by type	Media	Displacements times
	Population by age and sex	Number of cinemas, theatres, museums	Existence of general plan
	Population by level of studies	Number of gas stations	Forest area
	Population by nationality	Pharmacies	Hotel places
	Population by section	Policemen	Municipal surface
	Population density	Schools	Number of homes by type
	Time uses	Sports facilities	Unemployment

Figure 2.6 Structure for a minimum sociodemographic diagnosis.

Source: Own production.

social and cultural capital, you must also consider at a minimum the economic capital, patrimonial, natural and ultimately, the territorial capital.

In conclusion, the double perspective through which sociodemographic diagnoses can be understood has been revealed. One of them is restrictive and includes demographic phenomena and very basic social aspects. The other, on which we are betting here, is to expand and develop the social part of the analysis, making a broad and deep (sociological) use of the social dimension. This type of diagnosis is considered at the top in the decision-making processes of the regions involved, since it offers a high-resolution image of the social, political, economic and cultural reality. However, like any scientific knowledge, it is an always incomplete image that must be complemented by others. This is the main reason why it becomes a requirement to use qualitative techniques, which allow us to uncover arguments, speeches and reflections on the topics that are treated in each context.

Notes

1 It is usually considered that the population with these characteristics is the one between 16 and 64 years old, and it is generally understood that this is a low-dependency population, or as it is frequently found, as a "non-dependent population". Of course, there are always exceptions, and speaking of a human being as not dependent on other people does not conform to reality, at least from a certain sociological perspective.
2 VII Barómetro de Redes Sociales y Destinos Turísticos de la Comunitat Valenciana (INVAT.TUR)

References

Carratalá-Puertas, L., Ruiz-Callado, R., de-Gracia-Soriano, P., & Casasús-Chorques, A. (2018). Diagnóstico territorial. El litoral sur. In SERVEF (Ed.), *Territorio y empleo. Desarrollo territorial y mercado laboral valenciano Avalem Territori* (pp. 761–779). Valencia: Universitat de València.
Durán, M. Á. (2018). *La riqueza invisible del cuidado.* Valencia: Universitat de València.
Instituto Nacional de Estadística. (2017a). *DIRCE.* Retrieved from www.ine.es
Instituto Nacional de Estadística. (2017b). *INEBASE.* Retrieved from www.ine.es
Instituto Nacional de Estadística. (2017c). *Nomenclátor.* Recovered in www.ine.es
Instituto Valenciano de Estadística. (2017). *Banco de datos territorial.* Retrieved from www.pegv.gva.es

3 Exchange and hybridisation between disciplines in research and projection of healthy territories in the European Mediterranean basin

Carlos Arturo Puente Burgos,
María Jiménez-Delgado, Diana Jareño-Ruiz

Introduction

One of the benefits of interdisciplinarity is that it allows going beyond what one discipline alone can accomplish. In fact, each discipline carries out studies within its own field of knowledge that allow it to approach reality from one viewpoint. This viewpoint however, is limited to what it is able to address within the specific area of knowledge, while interdisciplinarity provides a shared structure in which to operate. There are many cases in which there is a risk that utilising the paradigms of a single discipline will end up generating unexpected and often counterproductive effects. This phenomenon is observed especially when working in the field of social problems, which tend to be complex in nature. While the intention is to generate positive contributions for the construction of public policies, these types of situations often lead to ambivalent results. Frequently, policies are correct from a conceptual point of view, but they do not solve the problem in question (Dunn, 1994).

The population settlements in the Mediterranean basin are an example of the way in which incorrect solutions have been given to particular situations. At times, these solutions were administered on the basis of similar habitat or through preconceived schemes and then they were transferred to very different environments. This inappropriate transference has violated territories with a great historical social, anthropological and cultural tradition. Urban planning schemes and land use were used in ways that were not in line with the Mediterranean lands' traditions. The possibilities offered by Mediterranean territories lie in the great diversity of geographic environments with relevant physical and geological characteristics. Said territories ought to continue with their own social and cultural development with the help of disciplinary professionals. In fact, disciplinary professionals who have entered into a process of re-education are best suited to address this in a comprehensive manner.

Understanding the complexity of the Mediterranean territories certainly requires disciplinary paradigms, however, disciplinary paradigms should not guide the particular projects and research on each territory alone. Whoever displays solid disciplinary knowledge armed with great humility marvels at the different realities. Their spirit of seeking knowledge will open up possibilities offered by other forms of knowledge, overcoming visions that pretend to lead to thinking that the welfare of people can be understood and solved through a unique disciplinary contribution:

> Who can take the measures necessary to the accomplishment of this task if not the architect who possesses a complete awareness of man, who has abandoned illusory designs, and who, judiciously adapting the means to the desired ends, will create an order that bears within it a poetry of its own?
>
> (Le Corbusier, 1989, p. 87)

This chapter seeks to provide concepts and reflections on interdisciplinary work to address complex problems such as the consolidation of healthy territories in the Mediterranean basin. The objective is to generate comprehensive solutions based on the hybridisation between disciplines. The observation and promotion of the territories as propitious for the health of its inhabitants and of those who reside in them for short stays are fundamental actions to be accomplished.

About interdisciplinarity

Even though at the beginning of the 19th century, science was considered as the integration of a series of disciplinary bodies with separate and specific knowledge, recognition was given to the fact that the knowledge involved in each discipline implied at the same time the reference to other disciplines and that the understanding of any particular portion of human life involved a wide range of fields. It began to be understood that a conception of science constituted by watertight compartments was a pernicious way of observing the world since its complexity was not exclusive to a single discipline (Crease, 2010).

In the '20s and '30s of the 20th century, the natural and formal sciences as well as epistemology generated a movement in favour of interdisciplinarity, with outstanding developments that gave rise to quantum physics, the theory of chaos, the theory of complexity and systems theory in the physical and biological sciences. Environmental studies became an example of an interdisciplinary subject aimed at transcending the divisions between the watertight disciplines. From a systemic approach, environmental studies considered the planet as a whole, where its parts interrelated with the consequent affectation of one another. For the social and human sciences, subjectivity, alterity and artistic and spiritual expression began to take on special validity. After the Second World War there was a porosity in the frontiers of knowledge, with emerging thoughts such as Ludwig von Bertalanffy's theory of systems, where he proposes that physical, biological, psychological, social and cultural phenomena allow the unity of the organic systems from the interrelation and interdependence between these elements (Uribe & Núñez, 2012).

The effects that could have on humanity due to the fragmentation of knowledge led to thinkers such as Edgar Morin, Basarab Nicolescu, Erich Jantsch and

Jean Piaget, who encouraged the Organisation for Cooperation and Development (OECD) to organise the conference "Interdisciplinarity, Problems of Teaching and Research in Universities" in the year 1979, in Nice, France. For their part, Morin and Nicolescu organised in 1994 in the Convent of Arrábida, Portugal, the first World Congress of Transdisciplinarity, where the concern for the hyperspecialisation of disciplines and subdisciplines and the effects that could be had on human nature was discussed. In response to this meeting, UNESCO raised the need for a reflection on the role of social and human sciences, for its significance in guaranteeing a humanist spirit that was already showing pernicious signs of regression in the world.

Such concerns led UNESCO to organise world events on the subject of interdisciplinarity between 1983 and 1985. Another event was organised with the International Center for Transdisciplinary Studies (CIRET) in 1997, where Edgar Morin, based on the knowledge of complexity, proposed to reform the university by the demands made by civilisation (Uribe & Núñez, 2012).

From then on, universities began to accommodate more interdisciplinary studies according to their own possibilities and regimes. Aware of their role in forming research groups that had thematic coherence, but that at the same time gave space to all fields of knowledge, universities thus began to make changes in their structures to include spaces in which research and interdisciplinary work could be hosted. The participation of academics from different disciplines, even from different countries, was included to facilitate the possibility of generating new thematic approaches. Thanks to the hybridisation between disciplines, new thematic approaches were generated that gave rise to innovative research fields based on the work of research groups involved in interdisciplinary projects.

However, the challenge of the growing differentiation between disciplines and the search for an encounter between sciences and humanities is still latent, which is why it is necessary to create spaces, both physical (Puente, 2012) and research-based, favourable to encourage, mediate and host interdisciplinary exchange. It is necessary to re-establish the unity between research and teaching as well as open borders between disciplines.

The 1990s saw the generation of multiple interdisciplinary research centres (Stahler & Tash, 1994); some were affiliated to universities whose purpose was to involve interdisciplinarity as part of their policies, where the meeting and dialogue of the disciplines would occur. The goal of these multiple interdisciplinary centres was to try and overcome the hermeticism of closed departments in their own specialised fields and create favourable conditions for professors from different fields of knowledge to develop complex research and projects the disciplinary approach could not address. It must be recognised, however, that in many cases interdisciplinary approaches were possible thanks to the initiative of professors who, bypassing the rigid organisational schemes of their universities, created spaces that integrated the participation of professors from different disciplines. Those professors knew that their own research or intervention in a specific reality or knowledge were not enough, but instead required other disciplinary visions to overcome the limitations found.

The interdisciplinary approach entails cooperation while facing a common project: preserving the value of each discipline. Preserving each discipline can be accomplished through dialogue with others and overcoming each discipline's limitations. Usually, interdisciplinary knowledge has to be confronted so that it responds to real needs and problems. An interdisciplinary conception of knowledge also implies an understanding of its complex and dynamic nature.

Appropriating the concept of limitations ("limes") by the philosopher Eugenio Trías (1999), crossing the limit is vital because what is beyond the discipline is necessary for its development and progress. Trías insists on the fecundity of crossing borders and settling in the limes. To cross the limit is to understand that there is a door and not a wall. Within this limitation one does not need to abandon a single discipline; interdisciplinarity is about putting one's discipline at the service of a deeper and more complex knowledge of the objects and problems of the world, from which the different disciplines can make an original and creative contribution.

This crossing borders by breaking walls or opening doors implies understanding the other and the others, as a cognitive but also political and moral act, and for this question dialogue is essential (Benhabib, 2002). Interdisciplinarity is not the juxtaposition of disciplines but the dialectic between them. This permeability does not mean the dissolution of methods and epistemological conceptions in a single discipline, but that the perspective is always open to extend their limits in a way that facilitates a deeper knowledge and a better response from science to human problems (Barry & Born, 2013).

According to Lefroy (2013), interdisciplinary research concerns, in addition to the conceptual and methodological issues, also the way in which researchers and citizens in general participate in it. Interdisciplinary work is understood as a way of investigating complex problems in order to find integral solutions.

The interdisciplinary approach to healthy territories

The co-evolutive process of interrelations and physical and environmental interactions in the Mediterranean territories has given rise to a particular history where, in the same way, the sociocultural components also co-evolve, interrelate and present a great diversity. That is why physical and nature studies of such territories must have a clear correspondence with the study and understanding of the societies that inhabit them. The effort to understand a social environment must take into account such interconnection, since the way people interrelate in their environments is complicated by the complexity of their environments, which in some way shape cultural expressions and the way they carry their forms of socialisation.

In consideration of the above, exploring the interdisciplinary dimension of the healthy study territory has four motivations: (1) its complexity, both natural and social, (2) the need to explore problems and questions of the territory that are not confined in a single disciplinary field, (3) the search for solutions to a need in their communities, and (4) having new technologies that provide special opportunities for the solution of problems (Frodeman, 2010).

An example of interdisciplinary work is found in the exercise developed for the consolidation of a healthy territory in the municipality of L'Alfàs del Pi, in Alicante, Spain, coordinated by the AEDIFICATIO Research Group of the University of Alicante. This integrated group witnessed the participation of several different departments: the Departments of Building and Urban Planning and Sociology I of the University of Alicante; Architecture and Urban Studies of the Politecnico of Milan; Communication of the Universidad Politècnica of Valencia; Architectural Construction of the University of Granada; Urban Policy and Planning of Hunter College, City University of New York; Sociology I of the Universidad Complutense of Madrid; and the Institute of Public Health of the Pontificia Universidad Javeriana of Bogotá. All these departments worked under a methodological interdisciplinary approach. This approach sought to improve the quality of the results from the methods contributed by the different disciplines to respond to the needs of the research or of the projects proposed around a healthy territory, in an exercise of continuous sharing (Klein, 2010).

This exercise required generating an appropriation of the languages, methods and corpus of each discipline so that the fundamental concepts can pass from one discipline to the other. The results achieved are a combination of adequate team member selection and appropriate coordination. This coordination has allowed for an interactive contribution of knowledge and a transcendence of the disciplinary limits to address the territorial problem that is the object of the work. This interdisciplinary teamwork has been achieved by overcoming the barriers of technical communication through a fluid and simple dialogue carried out without imposing points of view that seek to delegitimise knowledge from other sources, which also include popular and ancestral knowledge.

One of the premises that gave rise to the hybridisation of the different disciplines in the study of the territory of L'Alfàs del Pi was the recognition that the phenomenon to address transcended any of the disciplinary fields included in the exercise. This collaboration represented an even greater challenge for the achievement of interdisciplinary work, because diverse disciplinary fields were involved and methodological and even cultural paradigms specific to each international university and their social and political contexts were at play.

At present, the Mediterranean territories are faced with two realities that can clash with each other: the first, that of ancestral people who have developed their own social and cultural schemes, with their own ways of relating to their physical and natural environment, and second, that of new inhabitants, permanent or temporary, that bring with them new forms of behaviour against the environment they have inhabited. The former, exhibiting forms of life moulded by their environment through the centuries, the latter wanting to mould the environments from models brought from their previous contexts. This duality is how territories and social environments are overwhelmed by aggressive forms of urbanism and behaviours that end up being annoying for both categories of inhabitants.

To understand the Mediterranean territories, we must recognise the complexity of a phenomenon that day by day becomes increasingly complicated to analyse. This challenge makes it necessary to approach the complexity from new ways of understanding the physical, social, geographical, anthropological, cultural, historical, architectural and urbanistic components. It is, therefore, significant to achieve a new reinterpretation of these territories, as well as accept the complexity involved in their study, which is why this research is characterised by interdisciplinary work. This interdisciplinary work was conducted without neglecting the importance of disciplines that seek the understanding of singular and specific aspects that only each of them can address. In that sense, it is important to emphasise that good interdisciplinary work starts from the disciplinary strengths that each of the team members has.

In the foregoing, the public entities of each territory play a fundamental role that must combine the different ways of understanding the territory without losing sight of its sustainability and the quality of life of all those who inhabit it, including the right of everyone to enjoy a healthy environment and its scenic and natural qualities. In order to achieve this overall concept of health, the contributions of an interdisciplinary approach with its solutions and research must be taken into account.

Here, the responsibility of local authorities and the inhabitants of the territories as intermediaries to define the different actions is vital. Clear participation mechanisms where people are planners and executors of the different territorial plans should be put in place (Urdiales, 2003). There is another political fact of great relevance to acknowledge: there is a choice for Mediterranean territories between heteronomy (determination by another) and autonomy (self-determination) that would allow solving issues that concern the personal and local sphere.

The norms arising from the dialogue among all disciplines are re-thought, turning interdisciplinarity into a stronger and more cohesive type of study. This dialogue includes popular knowledge and traditions, as in the portrayals of all social agents and disciplines that come together to form the concept of interdisciplinarity, which must be identified with studies that involve the combination of two or more academic disciplines in one activity.

References

Barry, A., & Born, G. (2013). *Interdisciplinarity: Reconfigurations of the social and natural sciences*. London: Routledge.

Benhabib, S. (2002). *The claims of culture: Equality and diversity in the global era*. Princeton: Princeton University Press.

Crease, R. P. (2010). Physical sciences. In R. Frodeman (Ed.), *The Oxford handbook of interdisciplinarity* (pp. 79–87). Oxford: Oxford University Press.

Dunn, W. (1994). *Public policy analysis: An introduction*. Englewood Cliff: Prentice Hall.

Frodeman, R. (2010). *The Oxford handbook of interdisciplinarity*. Oxford: Oxford University Press.

Klein, J. (2010). A taxonomy of interdisciplinarity. In R. Frodeman (Ed.), *The Oxford handbook of interdisciplinarity* (pp. 15–30). Oxford: Oxford University Press.

Le Corbusier, C. E. (1989). *The Athens charter*. Retrieved from https://jasonsedar.files.wordpress.com/2011/03/the_athens_charter.pdf

Lefroy, T. (2013). Interdisciplinary research is about people as well as concepts and methods. In G. Bammer (Ed.), *Disciplining interdisciplinarity: Integration and implementation sciences for researching complex real-world problems* (pp. 365–374). Canberra: ANU E Press.

Puente Burgos, C. A. (2012). El Centro de Investigación Interdisciplinaria (ZiF) y la Universidad de Bielefeld. In M. C. Uribe (Ed.), *La interdisciplinariedad en la Universidad contemporánea. Reflexiones y estudios de caso* (pp. 492–509). Bogotá: Editorial Pontificia Universidad Javeriana.

Stahler, G. J., & Tash, W. R. (1994). Centers and institutes in the research university. *The Journal of Higher Education, 65*(5), 540–554. doi:10.1080/00221546.1994.11778519

Trías, E. (1999). *La razón fronteriza*. Barcelona: Destino.

Urdiales, M. E. (2003). Las cuevas-Vivienda en Andalucía: De infravivienda a vivienda de futuro. *Revista Electrónica de Geografía y Ciencias Sociales, 7*, 146(051). Retrieved from www.ub.edu/geocrit/sn/sn-146(051).htm

Uribe, M. C., & Núñez, J. (2012). Interdisciplinariedad y transdisciplinariedad: ¿colaboración o superación de disciplinas? In M. C. Uribe (Ed.), *La interdisciplinariedad en la Universidad contemporánea. Reflexiones y estudios de caso* (pp. 26–63). Bogotá: Editorial Pontificia Universidad Javeriana.

4 Preservation of tangible and intangible heritage in Mediterranean cities

Research tools and protection actions

Carlo Manfredi, Antonio Jiménez-Delgado,
Francesco Carlo Toso

The notion of cultural identity

The definition of cultural identity is being challenged by the return in the political scene, on an international level, of rising nationalism and the reinforcement of borders that had been blurred by the expanding notion of the global village firstly described by McLuhan (1964). An apparent wearing of the ideals of the Enlightenment in Europe and in western society accompanies the idea of cultural identity as a defence against the fear of an impoverishment of identity values due to global cultural homogeneity. National identity was, after all, the first incentive to the transformation of European society we experience today.

The geographically strategic position of the Mediterranean is the reason for its uniqueness, as literature refers to a "Mediterranean model" or "Mediterranean phenomenon" (Cassano and Zolo, 2007, p. 18), meaning the manifold of direct relations, peaceful or conflictive, that have been unfolding through history and still happen between the countries with a sea coast. Referring to this model, some observers acknowledged the existence of two other "Mediterranean Seas": one, sharing the coastlines of North, Central and South America is the Gulf of Mexico and the Caribbean Sea; the second is the South China Sea extending between China, South-East Asia and the South-Asian archipelagoes. These geographical unities, however, include a smaller number of nations and historically witnessed foreign cultural hegemonies, a situation largely different to the complexity of the Mediterranean, where cultures as diverse as European and Islamic meet. The Mediterranean is a "pluriverse" (Cassano, 2007, p. 84), within which different powers have been mixing and producing an ongoing layering of signs that are still visible today. Above all, the French historian Fernand Braudel acknowledged this complex unity, focusing his research on the topic of Mediterranean cultural identity. He specifically recognised one defining trait of this shared identity in the ways material culture adapted to different Mediterranean environments (Braudel & Duby, 1986; Braudel, 2009).

Identifying cultures allowed for their study and understanding through standardised types and categories, but these represent an outdated paradigm (Huntington, 1996). In fact, as the French philosopher François Jullien wants to prove, cultural identity does not exist (Jullien, 2016). Jullien's paradoxical thesis is based on an ample reflection: defending identity as the specificity and difference of a cultural model against another is unfruitful. Such a defensive attitude crystallises ideas, brings a uniformity of character and excludes every gradual divergence and comeback of ideas that bring productive innovations to a culture. One should rather talk about deflections, *écart*, between different cultures, and comparisons and contrasts between different approaches. Establishing contacts and interactions appears to be the only possibility to measure difference and change society. As was made clear by the development of language, society and cultures can only stay vital by evolving.

Architectural preservation: methodological issues

Reading the complex cultural layering we are discussing on historical urban centres is a complex research process, of which we'll outline the main lines of inquiry. The traces and signs that are left on the buildings and urban fabric are results of economic, political, military and religious reasons. Understanding the development of built heritage in its characteristics, and also those that are not evident, is the basis for understanding current awareness of heritage values and their future potential. Understanding the relations that bind single urban occurrences in a homogeneous urban fabric is at the basis of a coherent and aware architectural and urban planning. The aim is to investigate the material consistency of the artefacts, following the different methodologies that may provide information on the production and construction process, used materials and applied techniques. This kind of investigation lets us frame the history of a single urban settlement in the wider picture of long-term developments, in the tradition of the Annales school of historiographic thought.

In a first survey phase, the architectural researcher has a number of options and tools that need to be tailored to the specific needs and characteristics of the study object. Sources for a documental search need to be gathered, and archaeological tools are to be deployed to read the stratigraphic sequence of the studied built objects; suggestion for further instrumental analysis must be made when unclear results occur. This study path derives from different disciplines pertaining to historical construction research that include architectural history, construction history, the archaeology of standing buildings as well as historical research in the strict sense. The latter focuses on the sources of socio-economic history and allow us to gather a more general view, thus constituting a framework for more detailed analysis.

Starting with documental research, it must be noted that often its results will be uneven and incomplete. Only the major, most significant buildings may have left (mostly partial) documental sources on their management and maintenance. Aside from these fortunate findings, identifying sources on territorial management that have left long-standing traces will prove more fruitful and systematic, as shown by the studies conducted by Gallego Roca (1997) on the cadastral sources of the Kingdom of Granada.

Starting from these findings, an archaeological survey of the standing built heritage can be carried on. Archaeological methodologies have been developed in Europe over the last decades and are largely adopted in Spain too, where the academic debate of recent years has bridged the gap that had started since the 1930s, especially with the work of Javier Gallego Roca, Camilla Mileto and Fernando Lopez Vegas-Manzanares. The aim of such methodologies is to identify the constructional features of standing buildings, isolating especially those techniques and materials that had a more intense development in the studied territory. The techniques of archaeological analysis focus on the relations that building elements bear with each other according to the subsequent construction phases that took place in the lifetime of the building. It is, in the words of Mileto and Vegas (2003, p. 189), a methodology of stratigraphic analysis, born in the archaeological world and initially applied mainly to archaeological stratigraphy. It is not a simple reading, rather a proper analysis, so that the correct definition should be "construction stratigraphy analysis". The idea behind this approach is that the building can always be considered as a palimpsest bearing the traces of constructional events, sometimes even historical events in a broader sense. Analysing these events brings us to a diachronic understanding of the artefact: an understanding of how certain realisations, implemented materials and constructional techniques directly related to a particular past phase of the construction or maintenance of the building while hinting at the building's relations to the wider context. Thus the building itself becomes a theoretically unlimited source of knowledge, according to the different and sometimes discordant methodologies and lines of interpretation adopted. The consequence of this approach to studying historical buildings is that the possibility should be granted for future readings, analysis and interpretation. In the future, other interpretations, moving from analysis tools and study approaches that may be different to ours, should remain possible – and what matters most is that they will also be legitimate.

When moving from the analysis to the design of an intervention, the resulting plan should be respectful of the traces of the passage of time. The designer and planner should be aware that any new addition should only add new layers to existing ones, never replacing them, thus leaving the historical building as a text open to multiple interpretations. Furthermore, the analysis of the building as an artefact, when supported by the needed appropriate instruments, allows us to recognise issues connected with the decay of materials, with changes in geometrical and spatial configuration and with eventual structural instabilities. Eventually, the causes of damage can be identified, thus guiding appropriate planning and intervention.

The Mediterranean intangible heritage

The concept of heritage is complex and difficult to define as a development factor from a wide point of view. The development that the concept of heritage has undergone in recent years is well-known and brought us to the distinction between tangible and intangible heritage. On the one hand, new information technologies are a central tool for working on heritage and disseminating the results. On the other hand, heritage is inherited and represents not only a "scenery" to be preserved but also an economic factor. Nowadays the ability of the collectivity to maintain and increase its heritage has developed enormously. This ability to maintain heritage implies a higher responsibility for the management of intangible heritage and all those cultural manifestations associated with territory and population. One could therefore affirm that the cultural heritage of a nation or region not only consists of monuments and collections of objects in museums but also of living intangible or immaterial expressions of the local cultures, inherited from our ancestors and to be passed down to our descendants.

The 2003 UNESCO Convention on Intangible Cultural Heritage, resulting from a debate started in the 1970s (Sicard, 2008), recognised intangible heritage as an essential part of cultural heritage. According to the UNESCO definition, intangible cultural heritage consists of oral traditions and expressions (including language as a vehicle of heritage), performing arts, social practices, rituals and festive events, knowledge and practices concerning nature and the universe and traditional craftsmanship. Intangible cultural heritage is a complex heritage, which passes down knowledge and collective values and within which there are no restrictive definitions and instead often different cultural fields can overlap.

In the introduction to this volume, we address the existence of a "Mediterranean city" and a "Mediterranean culture" (Oltra 2000). Just as well, the concept of intangible heritage can be applied to a notion of "Mediterranean heritage", with its complexity and, at the same time, recognisability. An example of this dichotomy is the project "MEDINS. Identity Is Future: the Mediterranean Intangible Space", which aims at preserving, promoting and enhancing the intangible cultural heritage of the Mediterranean basin while strengthening the preservation and enhancement of the diversity of cultural expressions (de Caro, 2008). The project also proposed a cataloguing system for immaterial heritage.

Intangible Mediterranean heritage is part of the Mediterranean culture, placed at the same level of importance as tangible heritage, and not only from a cultural, economic and political point of view. The value of an Iberian amphora, a Greek temple or a painting by Sorolla is as relevant as, say, the stucco technique or Mudejar crafts; the folk tales and the songs by Spanish songwriter Serrat are as worthy of preservation as verses by Omero and Kavafis. It is an intangible heritage that presents a unique variety and a dynamic development, often shared by different Mediterranean countries; it is constituted of different "cultural heritages" that are enriched in their differences. In each one it is nevertheless possible to recognise elements of union and a common language. A clear example is the Mediterranean diet, declared by UNESCO to be intangible cultural heritage in 2010: "The Mediterranean diet constitutes a set of skills, knowledge, practices and traditions ranging from the landscape to the table. It is characterized by a nutritional model that has remained constant over time and space" (UNESCO, 2012, p. 20).

However, in order for the intangible heritage of the Mediterranean region to serve as a common language and bring its inhabitants closer, some preconditions are needed (Zouain, 2010, p. 187). First of all, the heritage must be acknowledged and accepted. Secondly, it must be promoted and at the same time, it must produce new values. In addition, it must be perceived, available and accessible. Finally, the values that this intangible heritage convey must be shared, at least in part. In this perspective, a fundamental role is played by education, in order for the Mediterranean inhabitants to know and promote their heritage and perceive it as an integral part of their own collective memory.

Evolution of heritage models and educational strategies

When discussing tangible and intangible heritage in Mediterranean cities, we can observe how models of heritage interpretation developed over time towards the inclusion of a more open, representative, plural and significant idea of heritage. In the context of the Mediterranean, the concept of heritage represents a particularly broad and strategic scenario to be used as a tool for local development. Some essential points which characterise the current heritage interpretation model are:

- It is open and includes all kinds of cultural manifestations coming from different groups, without establishing limited or closed criteria.
- It is representative, incorporating new and more numerous culturally relevant elements.
- Traditional cultural assets are considered an integral part of historical cultural assets. The sense of cultural continuity is emphasised. A heritage that is alive and in use is valued.
- The model is pluralist. The cultural witnesses of the ways of life, values and beliefs of different social groups are considered in their specificity and their role in society in general.
- It includes material and immaterial aspects, considering both dimensions as indissoluble.
- It has changing values over time. Heritage value are the result of a historical process. Heritage items are selected and endowed with content in a specific historical moment.
- It gives importance to socio-cultural contextualisation. Heritage can only be interpreted in its cultural context.
- In addition to movable and immovable heritage, cultural landscapes are included to be valued and protected.
- It includes cultural manifestations of specific groups, ethnic groups, ethnic minorities and nations. Preserving and promoting cultural heritage is a global and local issue, not just a matter of national interest.

In the framework of this education project, development and innovation projects, as well as educational institutions, promote initiatives to activate the heritage resources of the Mediterranean city. Educational strategies can consolidate around medium- and long-term policies, aimed at training and raising awareness of heritage resources in the Mediterranean area. A network which takes advantage of the educational value of Mediterranean cultural heritage would lead to designing and developing a wide range of activities with an extraordinary educational potential. Among the goals and tools of such a network are:

- to study and promote cultural tangible, intangible and landscape heritage through communication and education on a local as well as international level;
- to educate citizens on promoting and preserving their heritage;
- to connect local citizens with local heritage through different public and private organisations;
- to establish educational, artistic and cultural connections between different cities of the Mediterranean area;
- to produce multimedia information materials on arts and cultural heritage to be disseminated through the relevant media;
- to develop and implement courses, seminars, congresses, etc. with the aim of training cultural heritage professionals;
- to organise educational, artistic and scientific events with the involvement of professionals from other cities, regions and nations.

For the development of these aims, three stages are proposed with the following specific actions.
Study and research stage:

- preparation of documentation and multimedia materials for teaching programmes;
- publication of essays, articles and monographs not only for academic circles but also for the general public and primary education.

Production and promotion stage:

- creation of web pages and updating of the existing ones;
- preparation of didactic audiovisual materials and their diffusion on local and national TV through cultural programmes;
- publication of monographs about local artists;
- participation in international conferences;
- creation of headquarters in international organisations and participation in the European project;
- involvement of the press;
- full immersion in tourist circuits.

Heritage education stage:

- promotion of a network of universities to promote the awareness and protection of cultural heritage;
- promotion of coordination measures among cities for disseminating information and education on local heritage;
- coordinated actions through museums and local departments of education and communication;

- listings of cultural associations concerned with protection, dissemination and heritage education and creation of new ones;
- proposal for Master's degree or specialisation courses and networking of existing ones;
- design and development of workshops on education and heritage in non-formal contexts.

The contemporary cultural landscape of the Mediterranean area and its cities is of extraordinary cultural wealth, while paradoxically at the same time unjustifiable inequalities exist between populations on different Mediterranean shores. Therefore, one has to deal with an extremely complex space and extensive territories in the need for sustainable development initiatives. Culture in the widest sense, carrying universal values, is the primary tool for the design of actions serving the development of solutions that can make cities and territories prosper sustainably.

References

Braudel, F. (2009). *La Méditerranée: L'espace et l'histoire.* Paris: Flammarion.

Braudel, F., & Duby, G. (1986). *La Méditerranée, les hommes et l'héritage.* Paris: Flammarion.

Cassano, F. (2007). Necessità del Mediterraneo. In F. Cassano & D. Zolo (Eds.), *L'alternativa mediterranea* (p. 79). Milano: Feltrinelli.

Cassano, F., & Zolo, D. (2007). *L'alternativa mediterranea.* Milano: Feltrinelli.

De Caro, A. (2008). Il progetto MEDINS. Identity is future: The Mediterranean intangible space. In *El Patrimonio Cultural Inmaterial. Definición y sistemas de catalogación.* Actas del seminario internacional Murcia, 15–16 de febrero de 2007 (pp. 33–39). Murcia: Dirección General de Bellas Artes y Bienes Culturales.

Gallego Roca, F. J. (1987). *Morfología urbana de las poblaciones del reino de Granada a través del Catastro del Marqués de la Ensenada.* Granada: Diputación Provincial.

Huntington, S. P. (1996). *The clash of civilizations and the remaking of world order.* New York, NY: Simon & Schuster.

Jullien, F. (2016). *Il n'y a pas d'identité culturelle mais nous défendons les ressources d'une culture.* Paris: Éditions de l'Herne.

McLuhan, M. (1964). *Understanding media: The extensions of man.* New York, NY: McGraw-Hill.

Mileto, C., & Vegas, F. (2003). El Análisis estratigráfico constructivo como estudio previo al proyecto de restauración arquitectónica: metodología y aplicación. *Arqueología de la arquitectura, 2,* 189–196. Retreived from https://doi.org/10.3989/arq.arqt.2003.46

Oltra, B. (2000). A la luz del Mediterráneo. Culturas, civilizaciones y sociedades. *Barataria: Revista castellano-manchega de Ciencias Sociales, 2–3,* 13–26.

Sicard, H. (2008). Convención para la Salvaguardia del Patrimonio Cultural Inmaterial: conceptos e inventarios. In *El Patrimonio Cultural Inmaterial. Definición y sistemas de catalogación.* Actas del seminario internacional Murcia, 15–16 de febrero de 2007 (pp. 21–32). Murcia: Dirección General de Bellas Artes y Bienes Culturales.

Zouain, G. S. (2010). ¿Constituye el patrimonio cultural inmaterial un lenguaje común para el Mediterráneo? *Quaderns de la Mediterrània, 13,* 185–188.

5 Smart urban development in the Mediterranean city

An assessment framework

Raquel Pérez-delHoyo

Introduction

An urban model arises to solve the real problems and needs of cities and territories at a given time in history, and therefore pursues specific objectives according to the problems and the actual situation at the time in question. Thus, throughout history, very recognisable urban models have arisen, such as "The Expansion" or "The Garden City" models, to respond to the hygiene and mobility problems of industrial cities; or those stemming from "The Modern Movement", addressing housing and zoning problems. Obviously, for an urban model to be consolidated, the social, political and economic circumstances must be right (Vegara Gómez & De las Rivas, 2004).

The problems of today's cities are characterised by challenges stemming from the continuous growth of urban populations. Since the Industrial Revolution, cities have expanded significantly, with urbanisation playing an important role in increasing the value of cities and enhancing regional power. However, increasing urbanisation has had negative effects (Yin et al., 2015):

- Over-urbanisation puts great pressure on the city's aging infrastructure. It poses security risks. It leads to traffic congestion, problems with waste management. Urban blight widens the gap between rich and poor and all forms of social diversity.
- Over-urbanisation means overuse of resources. Cities in the European Union consume 70% of energy resources. Increased urbanisation has led to shortages of energy, water and land.
- Over-urbanisation turns into environmental degradation. Cities emit most of the carbon that damages the environment.

So what are the needs and objectives of today's city? They can be summarised in two areas: improving the liveability of cities and the quality of life of citizens and reducing carbon use and emissions. Can adopting a Smart City strategy make cities more efficient, reduce costs and ensure and improve a more sustainable quality of life? If the answer is affirmative, it is quite possible that we are faced with a new model of urban development, both current and future (Albino, Berardi, & Dangelico, 2015; Brdulak, 2015).

A Smart City strategy is a strategy that integrates all the services that a city needs by modernising as much as possible – according to the technological development – all the requirements for its public management. The objectives of good city management are also the objectives of a Smart City (Chourabi et al., 2012). Data collection, exchange and processing is vital for the management of an environment as complex as a city, therefore, information and communication technologies can make a substantial contribution to the way a city manages, innovates and solves its problems (Stimmel, 2016).

A Smart City is first and foremost a city that drives forward the quality of resource management and the provision of services using the latest technology and means. From this point of view, progress is being made towards an integrated understanding of the Smart City concept (Angelidou, 2014; Glasmeier & Christopherson, 2015). Therefore, Smart City projects should never be understood in isolation but as integrated elements in a city or region that constantly strives to improve its operation (Hollands, 2008). This way, the concept of Smart City belongs to a much broader concept and is part of the modernisation of the city (Höjer & Wangel, 2015).

Bearing in mind that the urbanisation process is continuing and that population density and consumption of resources in cities around the world are increasing, the beginning of any process of modernisation must be rooted in reflection on the following two questions: What kind of place do we want cities to be? And how should quality of life be defined (Murgante & Borruso, 2015)? Some other fundamental questions should be answered: Why and how will the city grow? What will the population structure be by age, and variety of professional fields? What type of medical services will be needed? What future energy costs, regional and international migratory movements or changes in the socio-economic structure of the population are expected?

Preferably, this discussion should be carried out by all stakeholders: municipal governments, citizens and businesses, supported by the experience of all relevant field experts, such as urban planners and architects, sociologists and psychologists and experts in services, technology and security. A large number of municipalities in the Spanish Mediterranean area are developing new Structural Land-Use Plans. The opportunity presented by the forthcoming approval of these Land-Use Plans makes the approach and discussion of these issues relevant (Gil-García, Pardo, & Nam, 2016). In this chapter, the case of the municipality of L'Alfàs del Pi is studied (Ayuntamiento de L´Alfàs del Pi, 2018a). L'Alfàs del Pi, like many other municipalities in the Spanish Mediterranean area, must take a stance regarding its future development model and establish its objectives, the realisation of which can be achieved with the support of technological solutions (Morandi, Rolando, & Di Vita, 2016).

Many organisations have created their own catalogues of criteria to define whether a city is smart or not (Figure 5.1). These criteria usually include all or some of the following categories: smart energy production and conservation, smart mobility, smart economy, smart living, information and communication

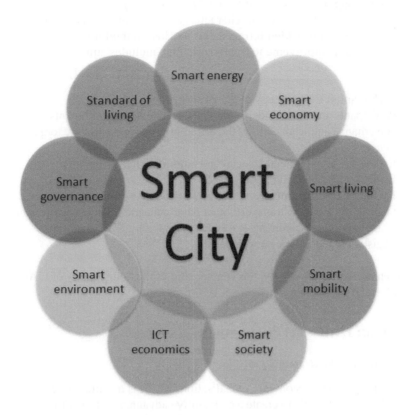

Figure 5.1 Smart City categories.

technologies (ICT) economics, smart environment, smart governance, standard of living and smart society.

In conclusion, Smart City is a new urban model based on knowledge. It is a new type of urban development based on in-depth research-exploration and wide application of the latest ICTs. The model is based on the transformation – modernisation – and development of the city and the convergence of innovative ICT applications. The model is vital for reducing carbon emissions and improving the livability of cities – in short, for the sustainable development of cities.

Objectives and methodology

Within the context of the project "Research on Coastal Cities: Anthropization and Atomization of Urban Models Towards a New Paradigm, Territory of Healthy Innovation and Urban Regeneration", the methodology used in the study *Comparative Study of Smart Cities in Europe and China 2014* (EU-China, 2016) is adapted to the characteristics and problems of the municipalities in the Spanish Mediterranean area. This methodology is based on an assessment and improvement framework that includes nine research-action axes:

- Global Smart City strategy of the city
- Participants and stakeholders – citizens, businesses, governments
- Governance and leadership of the different projects – structure, organisation
- Funding
- Economic, environmental, social and cultural value of the initiatives
- Business models that are being used
- Existing ICT infrastructure
- Smart City services: transport, open data, education, health, waste management, energy, environment
- Legislation and policies developed

As a representative case study, this chapter studies the municipality of L'Alfàs del Pi and its territorial context (Figure 5.2):

Diagnosis: L'Alfàs del Pi case study

Global Smart City strategy of the city

Within the context of economic globalisation, which is a context of competitiveness, cities must be able to create a competitive advantage for the different activities they carry out, not only economic but also cultural, residential, leisure and social. In order to create a competitive advantage, cities need to design a smart project for the future that highlights their uniqueness, i.e., the most unique and unrepeatable thing they can offer. That is why the keys to designing a Smart City project are found in the city's own context from a deep knowledge of the qualities that differentiate it from other cities.

In this sense, L'Alfàs del Pi has shown that it has a vision of the future and in identifying its vocation, which are signs of identity and the qualities that indicate its excellence in relation to its environment. L'Alfàs del Pi's ability to identify its vocation is expressed in the strategic documents drawn up and initiatives that were undertaken by the municipality in recent years.

The City Council of L'Alfàs del Pi has been working for some time on the elaboration of its Integrated Sustainable Urban Development Strategy (ISUD Strategy). In addition, with the slogan "Winter Sun L'Alfàs del Pi – L'Albir", the economic and social planning strategy of the municipality is determined. The ISUD Strategy aimed at promoting a tourism model of quality and healthy living supported by a productive economic model towards the service sector.

Figure 5.2 Smart municipalities in the Spanish Mediterranean area proposed.

The basic concepts that make up this ISUD Strategy are based on the components of excellence that differentiate L'Alfàs del Pi from other municipalities:

- Health – centres of healthy excellence, health centres, hospitals, polyclinics, residences, specialised medical clinics, rehabilitation, spa, relaxation and oriental medicine centres
- Sport – high-performance facilities, sports equipment, sports in nature, gymnasiums, athletics, competitions, tennis, football, golf, Nordic walking, basketball
- The environment – ravines, beaches, Mediterranean Sea, parks and gardens, agricultural areas, green areas, green infrastructure, Serra Gelada
- Social opportunities – associations, social centres, voluntary work, activities, day centres, accessibility
- Culture – monuments, schools, festivals, museums, cinema, music, theatre, parties
- Services – restaurants, markets, campsites, wineries, hotels, gastronomy

In addition, the municipality of L'Alfàs del Pi has demonstrated a conscious vision for the need of today's cities to position themselves in a competitive global context and to assume a new role. For this reason, it is important to establish strategic connections and networks with other cities based on common objectives, geographical location, interests, etc. In this sense, the initiative of the municipality to relate its descriptor to the context of the "European diagonal" and its strategic "diamonds" is coherent (Vegara & Ryser, 2008). The "City of Science and Innovation" distinction will also allow L'Alfàs del Pi to join the Innpulso Network, which is comprised of 62 Spanish cities and serves as a forum for cooperation and exchange of ideas and experiences.

Participants and stakeholders – citizens, businesses, governments

The municipal government is the project leader when it comes to analysing the groups involved in the decision-making process for the development of L'Alfàs del Pi as a Smart City. The municipality is currently in the process of developing its Structural Land-Use Plan, therefore, the regional government also collaborates directly in evaluating the strategy and objectives proposed by the municipality with a 20-year horizon.

In order to consider smart development as advanced a few elements need to be considered: the role of the entrepreneurial university and technology centres, potential land and housing developers, ICT service providers, system integrators, providers of other services in general and transport operators, etc. On the other hand, citizens' commitment is fundamental to the development of a Smart City. It is important to determine what role citizens play in the planning and design of the city and to involve citizens in the development and improvement of its services.

Many technologies can facilitate citizen participation, understood as collective thinking, such as crowdsourcing, collaborative social innovation, or gamification as a method of creating a playful environment for collaborating with citizens. The experience of public participation developed within the framework of the project "Research on Coastal Cities: Anthropization and Atomization of Urban Models Towards a New Paradigm, Territory of Healthy Innovation and Urban Regeneration" shows the commitment of citizens to the city. However, this participation must be encouraged and maintained through regular actions and by ensuring that these actions, with common objectives and interests, bring together the foreign and local populations.

Governance and leadership of the different projects – structure, organisation

The City Council of L'Alfàs del Pi has been developing a strong and competitive political leadership (Soler, 2016; Soler et al., 2016; Ayuntamiento de L´Alfàs del Pi, 2016–2018). Its political and administrative structure of governance is coherent and efficient. It resembles the structure of an intelligent city, setting in motion processes that allow interested groups to participate in decision-making with guarantees of transparency. Its web platform shows an appropriate use of ICTs to improve the overall governance of the municipality, activate innovation processes and enhance decision-making processes (Ayuntamiento de L´Alfàs del Pi, 2018b) (Figure 5.3). It is evident that a great effort is being made to modernise

Figure 5.3 Website of the City Council of L'Alfàs del Pi (www.lalfas.es/).

the system, and laying the foundations for a type of e-government. This effort in modernisation is resulting in important benefits for citizens, as well as the inter-connection of the different municipal offices, thus multiplying its effectiveness.

Through the municipal website, citizens are offered access to city informa-tion and institutional information. In addition, through the Internet, citizens can make enquiries and procedures in the virtual office, which is permanently open. This represents an important step forward in the relationship between citizens and the Administration, with significant improvement in bureaucratic proce-dures. However, although the intentions for modernisation and the entrepre-neurial will to lead are clear, it is necessary to provide the internal management units with sufficient personnel and means to achieve better and more effective governance. In a smart environment, people are the greatest resource for cities (Oliveira & Campolargo, 2015).

Funding

This section has not been evaluated, however, the enormous efforts that must be made in this regard are clear. For its part, the City Council of L'Alfàs del Pi has been working for some time on the preparation of its ISUD Strategy, a planning instrument that will allow the municipality to have European funding linked to the European Regional Development Fund (ERDF) for the period of 2014–2020.

Economic, environmental, social and cultural value of the initiatives

It is worth asking whether the development of the Smart City concept in L'Alfàs del Pi is serving to create added value to the economic, environmental, social and cultural activities taking place in the municipality (Dameri & Rosenthal-Sabroux, 2014). Are new businesses and jobs really being created, are CO_2 emissions, mobil-ity and traffic problems being reduced, are health-related services being improved, and how much benefit is being derived from the initiatives and investments made?

Both the ISUD Strategy, on which the City Council of L'Alfàs del Pi has been working for some time, and the initiative "Winter Sun L'Alfàs del Pi – L'Albir" (Soler et al., 2016) show that the actions carried out in the municipality are clearly oriented towards creating economic, environmental, social and cultural value. However, priority needs to be given to moving further in the direction of some of the proposed initiatives. In this sense, among the actions and projects planned by the City Council of L'Alfàs del Pi for the period 2016–2018, as described in the report that served this municipality for the award of the distinction of "City of Science and Innovation", are those classified in the Business section of the ISUD Strategy. These projects are intended to support business ventures and entrepre-neurs, with particular emphasis on creating innovative jobs, cutting-edge busi-nesses and attracting investments in contemporary initiatives. Innovation should take place in an optimal environment, and be aimed at improving the quality of life of citizens as well as responding to the real needs and demands of its popula-tion. Setting priorities in this direction seems important.

Business models that are being used

The initiatives undertaken, such as the L'Alfàs del Pi Urban Innovation Office or iL'Alfàs Intelligence & Innovation (Figure 5.4) make it clear that the municipality of L'Alfàs del Pi is at a medium-advanced level. However, these initiatives need to direct their efforts into concrete processes beyond a purely theoretical or strategic level.

Existing ICT infrastructure

This component is not considered to be a problem for meeting the challenges of today's cities. L'Alfàs del Pi has sufficient ICT infrastructure to undertake the initiatives that define its strategy. The Wi-Fi network is excellent. L'Alfàs del Pi's plan for the implementation of new technologies to improve services and attention to citizens is highly valued, specifically the plan for the implementation of Wi-Fi networks in public areas of the city.

However (and in line with Section 5), the attraction of innovative companies, investments and initiatives will undoubtedly favour the consolidation of the municipality's technological environment and its development towards a more intelligent model. This push for technological innovation is an opportunity to expand and consolidate the municipality's ICT infrastructure.

Smart City services: transport, open data, education, health, waste management, energy, environment

L'Alfàs del Pi bases its smart strategy on reconverting its territory towards a healthy socio-economic model of specialisation. The questions to be discussed are the following: What resources, specifically from the point of view of urban planning, are available to the municipality for the development of this strategy?

Figure 5.4 Website of iL'Alfàs Intelligence & Innovation (http://ilalfas.org/).

And how smart are these services and the way they are offered to the population? This section is probably where L'Alfàs del Pi can significantly improve its brand by considering working to improve the previously revised sections, and understanding that the various Smart City services are the result of permanent action-research, more innovative business models and an improvement in the municipality's ICT infrastructure.

Legislation and policies developed

The policies developed in relation to telecommunications, building construction, infrastructure management (electricity, gas, water), security and privacy, intellectual property, continuity of investment and business, etc., have a positive material impact on the smart development of the municipality. In general, the leadership exercised by the City Council of L'Alfàs del Pi is adequate to achieve its strategic objectives, and this is reflected in the different documents of its new Structural Land-Use Plan.

By way of conclusion of the diagnosis described, Table 5.1 shows a summary of the main characteristics common to Smart City projects and the brands that are taken in consideration in the municipality of L'Alfàs del Pi. The diagnosis has

Table 5.1 L'Alfàs del Pi: assessment framework

Characteristics	Maturity level					Not assessed
	Not addressed	Basic	Medium	Advanced	At the forefront	
Strategy					x	
Stakeholder				x		
Governance					x	
Funding						x
Value of the initiatives				x		
Business models			x			
ICT infrastructure			x			
Smart City services			x			
Legislation and policies developed				x		

1. Global Smart City strategy of the city
2. Participants and stakeholders – citizens, businesses, governments
3. Governance and leadership of the different projects – structure, organisation
4. Funding
5. Economic, environmental, social and cultural value of the initiatives
6. Business models that are being used
7. Existing ICT infrastructure
8. Smart City services: transport, open data, education, health, waste management, energy, environment.
9. Legislation and policies developed

been obtained according to the methodology used in the study *Comparative Study of Smart Cities in Europe and China 2014* (EU-China, 2016), and the corresponding assessment framework is established.

Proposed solutions

Suggestions for improvement of the advanced level components

With regard to Section 2, "Participants and stakeholders – citizens, businesses, governments", the involvement of the university is considered relevant as a focus of action-research in innovation processes. The university's involvement as an advisory body in the decision-making processes on matters affecting urban development in the municipality is also significant.

On the other hand, the participation of innovative companies and technology centres – universities – is essential to achieve the objectives of a Smart City strategy. In this sense, the provision of a physical space for its implementation, i.e., the provision of quality tertiary land, would specifically guarantee the link and commitment of these companies and centres with the objectives of the municipality. Soils adjacent to the axis of the national road number 332 – N-332 – could be the appropriate strategic space for this purpose. Furthermore, in a Smart City model, active communities capable of reaching consensus on future projects are necessary, and for this to happen, it is imperative to activate processes for the exchange of experiences and ideas. Citizen participation days will continue to be held on a regular basis, as it is being done in the municipality.

Regarding Section 5, "Economic, environmental, social and cultural value of the initiatives", the strategy of the municipality is focused on supporting business ventures and entrepreneurs, paying particular attention to the creation of innovative jobs and companies and the implementation of new initiatives. In this sense, including business innovation as an action strategy on the N-332 road is part of the municipality's strategic plan and may be an interesting option to assess. Several factors are in line with the municipality's territorial strategy for healthy innovation: generating new spaces capable of accommodating new uses of specialisation, promoting new innovation ecosystems on the N-332 road, and lastly making room for spaces suitable for experimentation and exchange as well cooperation and competition. This strategic innovation is committed to redefining its territorial model towards urban innovation, with the dual objective of improving the quality of life of its citizens and generating a new economy based on the specialisation of services linked to a healthy model of the territory. In fact, these new spaces – new technological companies, training, research, innovation and development centres – would complement the already existing spaces in the N-332 that comprise two centres of healthy excellence, a sports facility and a sports area in nature, two cultural centres, two river parks and eight shopping areas. These new additions would help the municipality to become a cluster of specialisation and reference in the field of health as a pole of attraction.

Finally, with regard to Section 9, "Legislation and policies developed", and in line with the previous comments, it should be pointed out that professional opportunities, an innovative environment, improved connectivity, a valued educational infrastructure, extensive residential options, quality of life, social balance, citizen security, varied cultural and leisure offerings, as well as the quality of urban spaces, are key factors in attracting both people and companies to L'Alfàs del Pi.

Proposals for improvement of the middle level components

With regard to Section 6, "Business models that are being used", it is necessary that the initiatives undertaken begin to concretize. In this sense, once again, action on the N-332 road can be an interesting option to assess. Generating a space that promotes a new innovation ecosystem, specialised in health and healthy living issues in accordance with the global strategy of the municipality, will undoubtedly attract interesting initiatives and new business models.

With regard to Section 7, "Existing ICT infrastructure", the attraction of innovative companies and investments in innovative initiatives would favour the consolidation of the technological environment of the municipality, being therefore an opportunity to expand and consolidate the ICT infrastructure of the municipality.

Finally, with regard to Section 8, it should be considered that "Smart City Services: transport, open data, education, health, waste management, energy, environment" are the result of permanent action-research, more innovative business models and an improvement in the ICT infrastructure of the municipality. It would thus seem that both a collaboration with the entrepreneurial university and an action plan on the N-332 road could be interesting options. In addition, citizens will need to receive information and training on how to use the new intelligent services along with the university's contribution.

Conclusions

From the study carried out in L'Alfàs del Pi and discussed in this article, two main points could constitute interesting options for the municipality. In the first place, a collaboration with the university should be implemented. The university could act as a means of action-research in innovation processes, as a service provider, as a training provider and as a consultative and advisory body in the decision-making processes on matters affecting urban development. The second main concluding point is that it is crucial to recognise the space occupied by the N-332 road axis as an opportunity to materialise the objectives of the municipality's global strategy. Overall, what the municipality of L'Alfàs del Pi should take in account is the primary focus of promoting a new ecosystem of innovation-oriented to the service and business sector, capable of welcoming new uses of specialisation, spaces for experimentation and exchange, cooperation and competition.

References

Albino, V., Berardi, U., & Dangelico, R. M. (2015). Smart cities: Definitions, dimensions, performance, and initiatives. *Journal of Urban Technology*, *22*(1), 3–21.
Angelidou, M. (2014). Smart city policies: A spatial approach. *Cities*, *41*, S3–S11.

Ayuntamiento de L´Alfàs del Pi. (2016–2018). *Memoria para la distinción "Ciudad de la Ciencia e Innovación"*.

Ayuntamiento de L´Alfàs del Pi. (2018a). *Plan General de Ordenación Urbana – Versión Preliminar*. Retrieved May 24, 2018, from www.lalfas.es/servicios/urbanismo/pgou-vsp/

Ayuntamiento de L´Alfàs del Pi. (2018b). *Sede electrónica*. Retrieved May 24, 2018, from www.lalfas.es/

Brdulak, A. (2015). Smart cities – a vision of the future or the present? In P. Golinska & A. Kawa (Eds.), *Technology management for sustainable production and logistics* (pp. 121–132). Berlin: Springer.

Chourabi, H., Nam, T., Walker, S., Gil-Garcia, J. R., Mellouli, S., Nahon, K., . . . Scholl, H. J. (2012). *Understanding smart cities: An integrative framework*. 45th Hawaii International Conference on Computer System Science (HICSS) (pp. 2289–2297). New York, NY: IEEE Press.

Dameri, R. P., & Rosenthal-Sabroux, C. (Eds.). (2014). *Smart city: How to create public and economic value with high technology in urban space*. Berlin: Springer.

EU-China. (2016). *China Academy of Information and Communications Technology, EU-China Policy Dialogues Support Facility II. Comparative Study of Smart Cities in Europe and China 2014. Ministry of Industry and Information Technology (MIIT), DG CNECT, EU Commission*. Springer.

Gil-García, J. R., Pardo, T. A., & Nam, T. (Eds). (2016). *Smarter as the new urban agenda: A comprehensive view of the 21st century city*. Berlin: Springer.

Glasmeier, A., & Christopherson, S. (2015). Thinking about smart cities. *Cambridge Journal of Regions, Economy and Society, 8*(1), 3–12.

Höjer, M., & Wangel, J. (2015). Smart sustainable cities: Definition and challenges. In L. M. Hilty & B. Aebischer (Eds.), *ICT innovations for sustainability* (pp. 333–349). Berlin: Springer.

Hollands, R. G. (2008). Will the real smart city please stand up? Intelligent, progressive or entrepreneurial? *City, 12*(3), 303–320.

iL'Alfàs Intelligence & Innovation. Retrieved May 24, 2018, from http://ilalfas.org/

Morandi, C., Rolando, A., & Di Vita, S. (Eds.). (2016). *From smart city to smart region: Digital services for an internet of places*. Berlin: Springer.

Murgante, B., & Borruso, G. (2015). Smart cities in a smart world. In S. T. Rassia & P. M. Pardalos (Eds.), *Future city architecture for optimal living* (pp. 13–35). Berlino: Springer.

Oliveira, A., & Campolargo, M. (2015). *From smart cities to human smart cities*. 48th Hawaii International Conference on System Sciences (HICSS) (pp. 2336–2344). New York, NY: IEEE Press.

Soler, V. (2016). *Territorio de innovación saludable L´Alfàs del Pí – Alicante*. Paper presented to II Congreso Ciudades Inteligentes.

Soler, V. et al. (2016). *"Winter Sun L´Alfàs del Pí – L'Albir – ". Estrategia de planificación territorial, económica y social*. L'Alfàs del Pi: Ajuntament L´Alfàs del Pí.

Stimmel, C. L. (2016). *Building smart cities: Analytics, ICT, and design thinking*. Boca Raton, FL: CRC Press.

Vegara Gómez, A., & De las Rivas, J. L. (2004). *Territorios inteligentes: nuevos horizontes del urbanismo*. Madrid: Fundación Metrópoli.

Vegara Gómez, A., & Ryser, J. (Ed.). (2008). *Building the European diagonal*. Madrid: Fundación Metrópoli.

Yin, C. T., Xiong, Z., Chen, H., Wang, J. Y., Cooper, D., & David, D. (2015). A literature survey on smart cities. *Science China Information Sciences, 58*(10), 1–18.

6 Exploring the (re)construction of healthy destinations in the Mediterranean

Carlos Arturo Puente Burgos, Rolando Enrique Peñaloza Quintero, Pablo de-Gracia-Soriano

Introduction

The consolidation of towns and cities as healthy territories must be the result of a series of political and social decisions that, accompanied by the technical work in a network of diverse actors (such as universities), seek to consolidate a series of actions aimed at the welfare and health of people who inhabit the territory. Consolidating towns as healthy territories also provides possibilities to visitors, including those seeking long-term stays (SIRHO, 2017).

The decisions taken by many municipalities to consolidate are diverse. Some municipalities are seeking to act on more than just a political declaration; they are operating along with the municipal administrations to restore territories' original characteristics. In fact, the Mediterranean territories' environmental landscape offers possibilities of enjoyment with leisure and relaxation alternatives, aspects that undoubtedly accompany good health.

Academia plays a prominent role in the consolidation of healthy territories by contributing to methodologies and techniques. Such techniques include developing a network with various territorial actors with the contributions of researchers and even foreigners. Researchers and foreigners carry out advanced work with the purpose of improving the health conditions of people in different territories. Examples such as L'Alfàs del Pi in Alicante, Spain, attest to this approach by consolidating an interdisciplinary work team coordinated by the AEDIFICATIO Research Group of the University of Alicante. The team is joined by members of the universities of Bogotá, Milan and New York. This partnership has allowed the execution of a workshop on Natural and Sociocultural Landscape which witnessed the broad participation of different social groups. Among these social groups, those in charge of the political decisions of the municipality attended. One of the main points that arose during the workshop, among other aspects, were interpretations of the objective of achieving a healthy territory. During the workshop several topics were discussed: the role that technologies would play in this process, the daily experience in the municipality and the main problems related to the environment and the health of people, as well as the possible solutions proposed and their different alternatives. Aspirations and ideas for the future were also discussed.

Particular idiosyncrasies of the territories located in the Mediterranean, with the coexistence of different nationalities, have become a focus. The aim is to enjoy good health conditions so that the territories' participation and their contributions can constitute an element of special importance. The goal in mind is to present such territories as healthy spaces. In order for a municipality to be considered a healthy territory, a series of elements and concepts linked to physical and mental health are necessary. In addition, the quality of the environment and the natural landscape, political decisions, social participation, health infrastructures, social and health services, housing and technologies are important factors for a territory to be considered a healthy space.

This chapter aims to provide a conceptual contribution so that future political, social and academic initiatives will have a precise foundation. These foundations will inform what should be taken into account for the purpose of providing health possibilities in the Mediterranean territories. These Mediterranean territories offer an environment conducive to greater wellbeing and better health for locals and visitors. Similarly, some initiatives that the municipal administration of L'Alfàs del Pi has been working on are presented. Proposals arising from the analysis of what was developed in the Natural and Sociocultural Landscape workshop are also presented. Both these initiatives seek to promote and strengthen people's overall health.

Key concepts

Health

Traditionally, health is considered the absence of disease. Of course, the previous statement is somewhat incomplete as it leaves out complexities about patients' physical condition that is to be determined by specialised health professionals. According to the World Health Organisation, "Health is a state of complete physical, mental and social wellbeing and not merely the absence of disease or infirmity" (OMS, 1946). This definition fully recognises that health interventions go beyond clinical services focused only on somatic aspects, but include also fundamental aspects of human beings such as psyche and connection to a community. The last two factors highlight human capacity for social relationships and imply therefore the importance of social relations and their interventions, which include interventions resulting from political decisions and those that influence people's health conditions (Alcántara, 2008).

The complexity of the health concept leads it to be somehow indeterminate. The health concept can refer to the different dimensions of a human being's health: social, political, economic, cultural, biopsychosocial, spiritual and environmental. Health is an aspect of the spiritual, mental, social and physical wellbeing of people that is of interest to different instances and social practices, such as politics and community life, that pursue the good life of a society (Faden & Shebaya, 2015).

Health, being the regulatory capacity that people have to achieve an adequate balance in their physical and psychic dimensions, is determined by biological laws common to all people. Health, however, is submitted to cultural norms of the epistemic world and other social practices and powers. In that sense, each person has their own individual health norm which is a result of their particular life history, personality and attachment to the environment (Granda, 2004). In this way, both health and illness are not only scientific categories but are also the result of policies, cultural practices and personal decisions, which are highly influenced by the social and political context (Navarro, 1997).

Julio Frenk (2003) works on those factors that determine people's health and expresses the following:

> Today we know that the health of a population depends on a multi-causal network of biological and social factors. The state of the environment, the forms of social organization, the economic structure, the level of democracy, the degree of urbanization, the material conditions of existence, schooling, nutrition, fertility and lifestyles are all crucial determinants of health, to which the health care system must respond. Health care is, therefore, an effort that goes far beyond the mere application of medical technologies.

During these factors' analysis, it becomes clear that the determinants of health are diverse, synthesised by Frenk in three general dimensions: basic, structural and proximal. Figure 6.1 presents a way to analyse such dimensions.

Public health

There are many approaches and movements around public health that have arisen at different times, but it wasn't until the second half of the 20th century that a significant shift was witnessed. It was particularly after the conclusions of the International Conference of Alma-Ata and the following conclusions of the WHO health promotion that views on health approaches broadened and people's well-being became more of a priority. Three aspects still continue to be the subject of scientific and academic discussions: (1) the political-economic organisation of societies, (2) their historical-cultural background and (3) the environmental conditions in which daily activities are developed (Mejía, 2013). Among the many definitions of public health, it is worth highlighting the Acheson Report (1988), which defines public health as "the science and art of preventing disease, prolonging life and promoting, protecting and improving health through the organised efforts of society". Health, as seen in a broader sense, helps us understand society by citizen groups, social movements and the State at its different levels.

In practical terms, the political and social effort is directed towards:

- The sanitation of the environment
- The control of communicable diseases
- The education of individuals in the principles of personal health and that of their community
- The organisation of medical and nursing services for the early diagnosis and preventive treatment of diseases
- The development of social mechanisms that ensure all people an adequate standard of living for the conservation of health; organising benefits in such a way that each individual is able to enjoy their natural right to health and longevity

BASIC DETERMINANTS

Population
Children, adults, seniors, nationals, foreigners

Environment
Drainage canyons, natural landscape, beaches, division of the territory, natural parks, urbanization, land use

Genome
Genetic material

Social organization
Economic structure, social participation, collectives, culture and ideology, traditions, government, technology and science

STRUCTURAL DETERMINANTS

Level of wealth
Poor population, family income by population according to location in the territory

Occupational structure
Level of employment, level of unemployment, economic activities, economic activities of nationals and foreigners

Social stratification
Division of the population into territories, inequities caused by the division of the territory

Redistribution mechanisms
Subsidies to the population, social services, income generation opportunities

PROXIMATE DETERMINANTS

Life conditions
Drinking water, sanitation, toilet, type and housing conditions by territory

Working conditions
Types of work, occupational health, remuneration of work, unemployment in homes

Life styles
Eating habits, nutrition, obtaining food, according to division of the territory, sport, free time

Systems of health
Access, distribution in the territory of the services, difference of opportunities according to territory

Figure 6.1 Determinants of health – analysis for L'Alfàs del Pi.
Source: Adapted from Frenk (2003).

A fundamental concept in public health concerns the social determinants of health: "The social determinants of health are the conditions in which people are born, grow, live, work and age. These circumstances are shaped by the distribution of money, power and resources at global, national and local levels".[1]

Health and environment

There are many definitions that have been given to the environment, but in this document it is considered "everything that is external to the human individual. It can be classified into physical, chemical, biological, social, cultural, etc., anything or everything that can influence the health condition of the population" (Yassi, Kjellström, de Kok, & Guidotti, 2002).

At present, it is already indisputable that human health ultimately depends on the capacity of a society to improve the interaction between human activities and the physical, chemical and biological environments (Figure 6.2). A balance must be struck between the safeguarding and promotion of human health and maintaining the integrity of the natural systems on which the environment depends. The physical and biological environments include everything from the immediate work environments to the home setting, regional and national habitat and also global environments.

The concept of environmental health, defined in 1993 at a WHO consultative meeting held in Sofia, Bulgaria, arises from the relationship between health and the environment: "Environmental health addresses all the physical, chemical, and biological factors external to a person, and all the related factors impacting behaviours. It encompasses the assessment and control of those environmental factors that can potentially affect health".

The 2030 Agenda for Sustainable Development (ECLAC, 2017), approved in September 2015 by the General Assembly of the United Nations, provides that the societies of each country generate their own transformative vision towards economic, social and environmental sustainability. The United Nations proposed 17 Sustainable Development Goals (SDGs) associated with the agenda. The agenda establishes that each country, and specifically the Mediterranean municipalities, must evaluate their own starting point and analyse and formulate the means to achieve this new vision of sustainable development embodied in the 2030 Agenda.

In this sense, the SDGs should be constituted as a planning and monitoring tool for the municipal territory. The SDG will constitute support for the path that will lead to sustainable development, which will be inclusive and in harmony with the environment. This is a task that municipalities must assume through public policies, budget, monitoring and evaluation instruments.

As stated in the prologue to the ECLAC document (2017): "The 2030 Agenda places people at its core and aims to achieve a rights-based sustainable development under a renewed global partnership, in which all countries participate at an

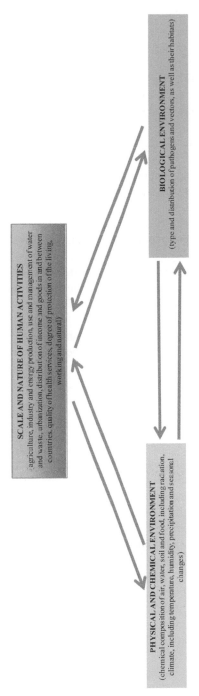

Figure 6.2 Interaction between human activities and the physical and chemical environment.

Source: Yassi et al., 2002.

equal footing". In particular, for the purposes of the health of the population, paying special attention to the following objectives is encouraged:

*   Goal 3: Ensure healthy lives and promote wellbeing for all at all ages
*   Goal 4: Ensure inclusive and equitable quality education and promote life-long learning opportunities for all
*   Goal 5: Achieve gender equality and empower all women and girls
*   Goal 6: Ensure availability and sustainable management of water and sanitation for all
*   Goal 7: Ensure access to affordable, reliable, sustainable and modern energy for all
*   Goal 8: Promote sustained, inclusive and sustainable economic growth, full and productive employment and decent work for all
*   Goal 9: Build resilient infrastructure, promote inclusive and sustainable industrialisation and foster innovation
*   Goal 11: Make cities and human settlements inclusive, safe, resilient and sustainable
*   Goal 12: Ensure sustainable consumption and production patterns
*   Goal 13: Take urgent action to combat climate change and its impacts
*   Goal 14: Conserve and sustainably use the oceans, seas and marine resources for sustainable development
*   Goal 15: Protect, restore and promote sustainable use of terrestrial ecosystems, sustainably manage forests, combat desertification, halt and reverse land degradation and halt biodiversity loss

Healthy territory

"Healthy territory" is understood to be the group made up of political and civil authorities, public and private institutions and organisations, businessmen and workers and the community in general, who in their daily lives seek to improve their living, working and cultural conditions. All the entities together seek to improve their living under conditions of a harmonious relationship with the physical and natural environment, but at the same time working to improve their coexistence and solidarity with democratic criteria. The ultimate goal is the promotion and maximisation of health, acting on the risks and determinants of health, seeking to improve the living conditions of the population in the different aspects that influence health conditions.

The territories of the Mediterranean basin are places of coexistence and life. They play a key role in the health of their inhabitants because it is in these territories that the living conditions of each person are determined. Such conditions are economic, social, cultural activities, educational or leisure. Equally, it is in the Mediterranean territories where housing, urban planning, traffic, etc. policies are decided. The territorial environment is where different forms of relationship and activity are carried out; the physical environment and natural support that has been adapted and modified by human beings determine the ways of life and use of time (leisure, work, obligations, needs) that can sometimes affect the health and wellbeing of people. People's wellbeing can be affected negatively as well, either because of social disintegration with groups that struggle individually for their own objectives or because there is a degradation of the environment. Thus, to develop an adequate health promotion, two fields are considered (University of Seville, s.f.):

- Social and collective, with the creation of an environment and favourable environments for the development of people, generating participatory and inclusive processes
- Individual, involving people, through the development of their own skills, so that they can adopt healthy lifestyles

The Healthy Cities project started in Europe in 1986 under the auspices of the WHO, which has allowed developing different alternatives in the countries aimed at promoting and protecting the health and wellbeing of citizens. This objective is pursued from a fundamental principle: the necessary interrelation of the various aspects that influence health and the necessary intervention of corresponding sectors, whether political, economic, cultural, social or environmental.

Experiences and uses of time

Generally speaking, the question of time in modern and advanced societies has been linked to an eco-political worldview which has been very focused on unlimited growth. The ideas of productivity and efficiency nowadays associate the concept of time as money, also known as "money-time" (Carrasco Bengoa, 2016). However, perhaps due to dispersion and social depth, different dimensions and perspectives seem to be hidden, taken for granted and even associated with attitudes and behaviours that could be labelled as pagan, abnormal and even deviant (Beriain & Sánchez de la Yncera, 2010). In this sense, social acceleration is a characteristic (Rosa, 2016) seen as a positive value and with little criticism of our societies. Technological acceleration, the acceleration of social change and the acceleration of the rhythm of life are three components that configure what is known as social acceleration, through, respectively, economy, social structure and culture (Figure 6.3).

People, and therefore organisations, have been socialised and accustomed to the ticking of the clock and the imposition of this type of time on everyday life. It has been necessary to wait until certain disciplines with a scientific basis begin talking about the problems derived from both the uses and the experiences of time, or modern use of time's repercussions on individual and collective health. A new perspective is needed to witness an opening to this new knowledge, widely recognised, but not yet known or learned by many. A few people are surprised to hear about distress (colloquially called stress) that work, multitasking and cities produce. The most obvious evidence is, perhaps, that despite the daily nature of these arguments, these issues are not incorporated into most countries' political

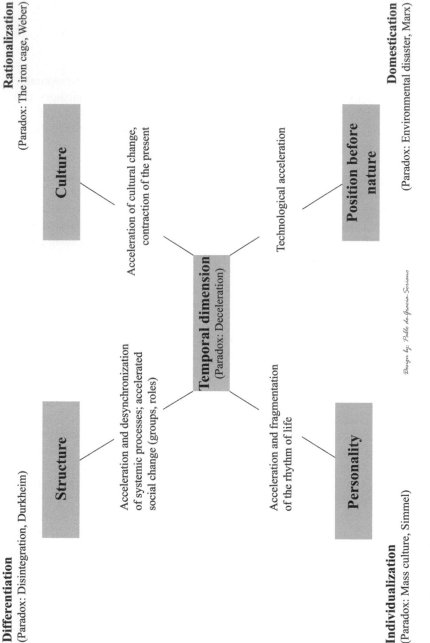

Differentiation
(Paradox: Disintegration, Durkheim)

Rationalization
(Paradox: The iron cage, Weber)

Structure

Culture

Acceleration and desynchronization of systemic processes; accelerated social change (groups, roles)

Acceleration of cultural change, contraction of the present

Temporal dimension
(Paradox: Deceleration)

Technological acceleration

Acceleration and fragmentation of the rhythm of life

Personality

Position before nature

Individualization
(Paradox: Mass culture, Simmel)

Domestication
(Paradox: Environmental disaster, Marx)

Design by: Pablo de-Gracia-Soriano

Figure 6.3 The modernisation process I and II.

Source: Translated from Rosa, 2011, edited and redesigned by Pablo de-Gracia-Soriano.

agendas and structures. Sociology, and particularly the sociology of time, prove this last point (Ramos, 2008).

The forms of organisation and the rhythms that mark social and productive systems appear as potential lifestyle sources. Current productive systems, however, entail high levels of stress and distress. Among others, stress is a relevant aspect when talking about health. The way social and productive rhythms are managed will give an account of the population's daily wellbeing, and therefore the quality of life.

There are alternative philosophies with different discourses that contrast the increasingly fast and stressful pace of life. One of them is a slow philosophy, materialised in the self-denominated slow movement. The slow movement shows a commitment to reduce the rushed characteristics of modern and postmodern global societies. Slow Food, for example, defends the idea that modern society moves more quickly than it needs to, and that it is necessary to know when to stop or slow down. In other words, the Slow Food movement arises from concerns about how intensely, in terms of acceleration and hurry, we live and organise our lives, our politics, our market-space and our work. Overall the Slow Food movement is concerned with the consequences generated by the dominant and increasingly fast rhythm of life.

Since the emergence of its first organised form, Slow Food International, founded in Italy by the journalist Carlo Petrini in 1986 (see www.slowfood.com), has established the foundations of philosophical currents and related practices (Novo, 2010) that pursue the search for "just" times (Honoré, 2005). Reunited under the title of slow movement, Slow Food is an organisational, self-reflective network (Castells, 2015), characterised by a reflexive identity construction created in juxtaposition to fast logic (Fernández Aguinaco, 2014).

Currently, it is known that altering rest time has dire consequences for our mental and physical health, both in the short- and long-term. And it isn't just about rest time. The rhythm of life that we have, conditioned by the hegemony of an accelerated socio-temporal model, produces stresses that range from depressive states to fatal disorders. In addition, such rhythms impose a use of space that tends to accelerate time's exploitation and, consequently, to produce certain environmental problems. The Industrial Revolution is an obvious example.

The foregoing leads us to consider how necessary it is, in the present and in the future, to take into account what social agents say or assume about time (Ramos Torre, 2007). It is crucial to rethink our approach to time both to direct our actions towards healthy models and scenarios and to reduce the risk of individual, social and environmental illness.

Proposed initiatives

As an example, some of the aspects that the municipality of L'Alfàs del Pi has worked on are introduced. Such aspects are oriented towards the achievement of a favourable territory for the health of its inhabitants and those who visit it. These aspects have been worked on by municipal officials, together with diverse groups, especially by those who participated in the Natural and Sociocultural Landscape workshop developed by the University of Alicante and its AEDIFICA-TIO Research Group.

Environment and health

For more than 50 years, the territory of L'Alfàs del Pi has been recognised by people of different nationalities, especially from Northern Europe, to have optimal qualities for the enjoyment of its landscapes, leisure and tranquillity. L'Alfàs del Pi is also known for its incomparable environmental characteristics which are all conducive to good health. Landscape and environmental quality constitute a key factor in the promotion of good health, both physical and mental, but also social.

Health and society

A healthy environment must have infrastructure that generates appropriate conditions for health, through the physical, economic and social regeneration of the environment.

Health and quality of the environment

The intimate relationship between health and the environment has clear effects on people if we consider that all the elements composing the environment, such as air, water, the work surroundings, homes interiors and buildings, have clear implications for health. Therefore, it is important to take care of the health and sanitation of the environment. We seek to generate a good-quality environment and those elements that contribute to it.

Health and information technologies

Within the framework of the Smart City concept, the application of information and communication technologies (ICT) is aimed at providing an infrastructure that guarantees citizen participation. Citizens must be informed of their decisions and must be educated about their own health. When citizens are aware of healthier choices, they can contribute to sustainable development, improve their own quality of life, utilise their own natural resources more effectively and be better participants in their civic duties.

Health and time policies

For years, problems known as "conciliation policies" have occupied the political agendas of several countries. The Spanish case is one of them. Conciliation policies have paid special attention to the double burden of women and also to the redefinition of the rights and duties of those who have sons or daughters, whether men or women. Similarly, they have also considered other problems that have been emanating from theoretical-empirical research, such as the case of "doppia presenza". This concept refers to being mentally in more than one place at the same time; that is, having thoughts at a given moment, which refer to physical or social spaces different from the one you are now (Balbo, 1994).

Space is not always enough to explain reality (Arbaci, 2007). Beyond this, it is convenient to distinguish between physical space and social space. One of the aspects that for decades has occupied part of academic sociology and certain political areas is the question of time and its experience, both in people's daily life and at structural and institutional levels. In Spain, we find some city councils that have decided to incorporate structures known as time policies into their agenda. Time policies not only work on issues related to temporary organisation, but also deal with problems ranging from inequality to the bases of welfare and day-to-day wellbeing. In this context, the proposals that are offered to the municipalities to work on the relationship between health and time are two, very general, although necessary:

- Observation of Time: Uses and Experiences (OTUE): It is important to know social uses of time in life, work and cities (Torns, 2011). Having this information means being able to conduct better municipal management to detect key aspects in the health field, although social use of time can be used to draw conclusions in other fields as well. The creation of an OTUE would suppose the production of data and information on the times observing people's lives. Such data would be catalogued and put to the service of the municipality, of citizens, of the decision-making processes and of investigations in this field. As an example, we can observe the Survey of Time Budgets (EUSTAT-EPT) developed in Basque Country.
- Time policies for daily wellbeing: The city, in addition to being an architectural ensemble, represents the relational space where people live and produce society as well as the place where people are closer to the rights and duties of citizens. In this sense, it is important to point out the possibility of local governments to know the social reality of their municipality and contribute to the reduction of some inequalities closely related to uses and experiences of time. Some of these challenging uses of time can explain mental and physical health problems. For example, two cases known for incorporating such time policies in the political agendas are the Council of New Social Uses of Time of the Barcelona City Council (Spain), and those cities spread across the Mediterranean and Europe which have opted to take advantage of the so-called slow cities movement, where space is redefined favouring certain practices over others. Those cities serve as an example of the pedestrianisation of certain streets (favouring social interactions) and promoting the use of non-polluting transportation (electric cars or bicycles). A valuable strategic source for the design and planning of these temporary policies for a healthy municipality is the OTUE.

These elements are increasingly becoming central to research on wellbeing (in a broad and structural sense) and daily wellbeing (individuals and their daily relationships with their lives, jobs and vital environments such as the city). Those three elements are search for balance in the three times mentioned above, understanding the different uses (quantitative and qualitative) that people make of time and the inequalities that are generated from use of time (Oso, Sáiz-López, & Cortés, 2017; Revilla, Martín, & de Castro, 2017; Prieto, Ramos, & Callejo, 2008). The Mediterranean and its cities exhibit changing characters that are continuously rediscovered (Oltra & Martín de los Santos, 2005) and they are based on tourism in the social and economic realms (Mantecón, 2017). Because of these qualities, Mediterranean dwellings require systematic observations that consider their dynamics and complexities, to make it possible for municipalities (physically and socially) to reflect on their health status.

Note

1   www.who.int/social_determinants/sdh_definition/en/

References

Acheson, D. (1988). *Public health in England: The report of the committee of inquiry into the future development to the public health function.* London: HMSO.

Alcántara, G. (2008). La definición de salud de la Organización Mundial de la Salud y la interdisciplinariedad. *Sapiens, Revista Universitaria de Investigación, 1*(9), 93–107. Recuperado de https://dialnet.unirioja.es/descarga/articulo/2781925.pdf

Arbaci, S. (2007). Ethnic segregation, housing systems and welfare regimes in Europe. *International Journal of Housing Policy, 7*(4), 401–433.

Balbo, L. (1994). La doble presencia. In C. Borderías Mondejar, C. Carrasco Bengoa, & C. Alemany (Eds.), *Las mujeres y el trabajo: rupturas conceptuales* (pp. 503–514). Barcelona: Icaria.

Beriain, J., & Sánchez de la Yncera, I. (2010). *Sagrado/Profano. Nuevos desafíos al proyecto de la modernidad.* Madrid: Centro de Investigaciones Sociológicas.

Carrasco Bengoa, C. (2016). El tiempo más allá del reloj: las encuestas de uso del tiempo revisitadas. *Cuadernos de Relaciones Laborales, 34*(2), 357–383.

Castells, M. (2015). *Redes de indignación y esperanza.* Madrid: Alianza Editorial.

ECLAC. (2017). *Agenda 2030 y los Objetivos de Desarrollo Sostenible. Una oportunidad para América Latina y el Caribe.* Santiago de Chile: Naciones Unidas.

Faden, R. y Shebaya, S. (2015). Public health ethics. *The Stanford Encyclopedia of Philosophy* (Spring 2015 Edition). Recuperado de http://plato.stanford.edu/archives/spr2015/entries/publichealth-ethics/

Fernández Aguinaco, V. (2014). Contra la prisa. El movimiento slow y derivadas. *Crítica*, (990), 80–83.

Frenk, J. (2003). *La salud de la población: Hacia una nueva salud pública.* México: FCE, SEP, CONACyT.

Granda, E. (2004). ¿A qué llamamos salud colectiva, hoy? *Rev Cubana Salud Pública, 30*. Ciudad de La Habana abr.-jun. 2004. Recuperado de http://scielo.sld.cu/scielo.php?script=sci_arttext&pid=S0864-34662004000200009

Honoré, C. (2005). *Elogio de la lentitud.* Barcelona: RBA.

Mantecón, A. (2017). El turismo residencial no existe. Revisión de un concepto y crítica de su función ideológica. *Cuadernos de Turismo*, (40), 121–144.

Mejía, L. M. (2013). Los Determinantes Sociales de la Salud: base teórica de la salud pública. *Rev. Fac. Nac. Salud Pública, 31*(suppl 1), S28–S36.

Navarro, V. (1997). Concepto actual de la Salud Pública. In N. A. Martínez (Ed.), *Salud Pública* (págs.49–55). Madrid: MacGrawHill Interamericana. Capítulo 3 recuperado de www.facmed.unam.mx/deptos/salud/censenanza/spi/fundamentos/navarro.pdf

Novo, M. (2010). *Despacio, despacio. . .* (2ª ed.). Barcelona: Obelisco.

Oltra y Martín de los Santos, B. (2005). El Mediterráneo "sub especie temporis". *Sociedad y utopía: Revista de ciencias sociales,* 121–144.

OMS. (1946). *Summary report on proceedings minutes and final acts of the international health conference held in New York from 19 June to 22 July 1946.* Official Records of the World Health Organization No. 2. Recuperado de http://apps.who.int/iris/bitstream/10665/85573/1/Official_record2_eng.pdf

Oso, L., Sáiz-López, A., & Cortés, A. (2017). Migraciones y movilidad social: Escalando la jerarquía social en el espacio transnacional. *RES, 3*(26), 293–306.

Prieto, C., Ramos, R., & Callejo, J. (2008). *Nuevos tiempos del trabajo* (p. 255). Madrid: Centro de Investigaciones Sociológicas – Colección Monografías.

Ramos Torre, R. (2007). Time's social metaphors: An empirical research. *Time & Society, 16*(2–3), 157–187.

Ramos Torre, R. (2008). Prolegómenos a una sociología del tiempo. *CIS, Reis: Revista Española de Investigaciones Sociológicas,* (122), 183–185.

Revilla, J. C., Martín, P., & de Castro, C. (2017). The reconstruction of resilience as a social and collective phenomenon: Poverty and coping capacity during the economic crisis. *European Societies, 20*(2), 1–22.

Rosa, H. (2011). Aceleración social: consecuencias éticas y políticas de una sociedad de alta velocidad desincronicada. *Persona y Sociedad, 25*(1), 9–49.

Rosa, H. (2016). *Alienación y aceleración. Hacia una teoría crítica de la temporalidad en la modernidad tardía.* Buenos Aires: Katz.

SIRHO. Instituto de Investigaciones Turísticas. (2017). *L'Alfàs del Pi, Destino Saludable.* Alicante: Universidad de Alicante.

Torns Martín, T. (2011). Conciliación de la vida laboral y familiar o corresponsabilidad: ¿el mismo discurso? *Revista Interdisciplinar de Estudios de Género (RIDEG),* (1), 5–13.

University of Seville. (s.f.). *Estilos de vida y Promoción de la Salud: material didáctico.* Recuperado de http://grupo.us.es/estudiohbsc/images/pdf/formacion/tema7.pdf

Yassi, A., Kjellström, T., de Kok, T., y Guidotti, T. (2002). *Salud ambiental básica.* México: Programa de las Naciones Unidas para el Medio Ambiente. Recuperado de www.pnuma.org/educamb/documentos/salud_ambiental_basica.pdf

7 Citizen participation in natural and sociocultural landscapes
Methodology and main results

Diana Jareño-Ruiz, Pablo de-Gracia-Soriano, María Jiménez-Delgado

Introduction: participation and citizenship

Scientific and academic agents and the majority of organisations and institutions recognise the relevance of citizen participation when preparing public policies. The international framework knows that well-constructed and maintained relations between the State and civil society generate stable structures for the future, in which the different interests of the communities are collected. This is the reason why it is considered very useful to explain the methodology and the different techniques used in participatory action research (PAR). In order for readers to be able to fully embody this methodology, they need to know the meaning that the authors attribute to the concepts of participation and citizenship.

The concept of *participation* used in this research is provided by Charles Wright Mills as "a permanent process of opinion formation, within working groups and intermediate bodies, around all the problems of common interest, as that these arise and require solutions, that is, decisions" (Riva, 1994, p. 30). This definition is based on reflection, which is the first step to reach the real and effective capacity of an individual or group to make decisions about matters that directly or indirectly affect their activities in society (Gyarmati, 1987). For civil society to be properly involved, it is necessary to integrate participation mechanisms within the same implementation processes, that is, as part of the programming of objectives, activities and tasks that seek to achieve a common goal (Martínez, 2007; Morgado, 2013). The importance of considering the actors in the design, implementation and evaluation of public policies is that

> social participation in civil society determines and thus favors the emergence of public policies and contributes to the organization of society in order to achieve the satisfaction of their needs and interests, in such a way that the action of these organizations has increased the participatory level of the inhabitants; in many cases, the organization of civil society aims to promote democratic values in their current practice and also to the way in which decisions are made.
>
> (UNDP, 2004)

For these reasons, the inclusion and democratisation of the participation processes can improve the design of social policies. Through citizen involvement, it is the people who can account for their needs and their possible solutions. Citizens shift from being passive subjects to becoming active players in decisions regarding environmental, climatic, demographic, social and economic dimensions of their municipality.[1]

Citizenship is the second concept to address and has its origins in classical Greece and, more recently, in the French Revolution. The term was identified with the idea that citizenship constituted freedom to participate in the power of the State. The interaction between both citizen and State allows us to understand the relationship between those who execute public policy and those who receive it, since it is possible to know the degree of involvement of citizens depending on the type of policy (Jiménez, 2008). As stated in the texts by Carlos Sojo (2002) and Jennifer Morgado (2013), it is interesting to analyse the conceptions of citizenship from a sociological perspective through Marshall, Habermas, Touraine and Canclini. The contributions of these last two authors, Touraine and Canclini, will be the basis of the concept of citizenship used in this work. For Touraine, citizenship refers to the free and voluntary construction of a social organisation that combines the unity of the law with the diversity of interests and respect for fundamental rights, that is, the construction of an eminently political space, not state or commercial (Touraine, 1995). Clanclini goes a step further, and he stops seeing the citizen only as a subject of rights granted by the State to conceive citizenship along with the cultural and social practices that create the sense of urban difference and, more specifically, of identity (Canclini, 1995). The revisions of these works finally emphasise the conception of citizenship as identity, which Canclini already defined: "citizenship and rights do not only speak of the formal structure of a society, they also indicate the state of the struggle for the recognition of others as subjects of valid interests, relevant values and legitimate demands" (Sojo, 2002, p. 21).

Thus, attending to the different edges that make up the previous concepts (participation and citizenship), *citizen participation* needs to be contextualised in order to be understood. Citizen participation is located in a space and time in which there is public power which has the intention on the one hand of expressing the general interest of a social unit and, on the other hand, of carrying out an administration oriented towards public power and improving its functioning (Baño, 1998). It is because of this double aspect that different typologies of citizen participation will appear; scientific literature distinguishes between different types: informative, advisory and proactive, resolutive, executing or controlling (Cerritos & Rodríguez, 2005). This project was proposed to make an advisory and proactive participation, which "alludes to the possibility that the government has to consult problems or own initiatives with the citizenship, as well as with the right that the citizens have to make proposals so that these are considered by the elected authorities" (Cerritos & Rodríguez, 2005, p. 14). Listening to the opinions of citizens does not oblige the authorities to obey them, but does enrich government decision-making and limit the risks (Morgado, 2013). Therefore, the idea is that the leaders of the different territorial areas consult, listen and agree with the citizens of the different territories on the best process for making socio-spatial decisions. Thus, PAR has become the key methodological strategy to achieve the participation of the population as a whole in public and private decision-making that affects all spheres of daily life.

Methodological framework for PAR

The complexity of social realities requires transcending the traditional fragmented conceptions of methodological approaches. The task of

> cutting back, dividing and typifying social reality prevents, if care is not taken, to capture the complexity of the population's problems and demands, and in the best of cases, traditional approaches allow us only to synthesize and explain (but in sometimes difficult to understand) the paradoxes of social life.
>
> (Francés, Alaminos, Penalva, & Santacreu, 2015, p. 11)

There is therefore no single social reality, but as many realities as subjects. This goes from the logical principle to the dialogical principle in the research process, based on abduction, which consists of adding discourses with which to project new possible scenarios (Montañés, 2009). In the words of Jesús Ibáñez (1985), one would be faced with the dialectical perspective in which the use of language and its capacity to register and trigger processes of action prevail. In the dialectical perspective, important changes take place, especially in the researcher's function and in the process of knowledge construction. From this perspective, the study subjects have cognitive and practical autonomy to produce information and act according to it (Francés et al., 2015). Discursive methodology and PAR techniques emerge from this dialogical approach.

The dialectical perspective conceives the research activity as a result of a self-reflexive process of knowledge, in which the subject is formed by all the members as objects of study, and in which the activity of the researcher is oriented towards producing knowledge with adaptive purposes (Valero, 2005). That is, the research process does not end with the analysis of results, but with its strategic application. Thus, PAR stands as the most efficient method to follow for the creation of positive social change (Seymour & Hughes, 2000).

The PAR method establishes distances with respect to a classic approach from the beginning of the research because the establishment of the demand and the research objectives are the result of a process of triangulation of actors. Triangulation can be defined by the opening of a negotiation space in which promoters, researchers and subjects of the community agree on the objectives and scope that it must have; it is then when the population, with the help of the research team, becomes an analyst of their own reality (Francés et al., 2015). The objective with the use of this strategy is to obtain valid knowledge that allows the team to satisfy the needs, reach the goals or pose new challenges in the researched groups.

Table 7.1 Main techniques of participatory research for social creativity and phases of research*

Stage	Techniques
Negotiation of the demand and participative construction of the project	Focus group Sociogram Cognitive map Participant observation
Self-diagnosis	Interviews and discussion groups Biographical techniques Photovoice SWOT Participatory survey Deliberative survey Jury citizen
Programming and implementation of actions	Problem and solutions tree Situational flowchart Workshop for the future EASW

Source: Translated from Francés et al. (2015, p. 91).

*Note: In Francés et al. (2015), each of the techniques for collecting information following the PAR strategy described in the previous figure can be consulted more in detail.

To achieve this goal, it is necessary to utilise quality research design in which the best techniques for collecting information are selected in accordance with the objectives set in each case, thus constructing dialogical and plural processes with transformative capacity. Participatory research techniques share a series of common characteristics: (1) they are techniques open to unscheduled information; (2) they assume in their approach a tacit symmetry and horizontality; (3) they work from second-order elections (consensus); (4) they have a strategic vocation; (5) they are applied in group contexts; and, (6) they are linked to return processes (Francés et al., 2015, pp. 85–88). In Table 7.1, the classification of the techniques according to the phases of the investigation is shown. These techniques become tools at the service of the community that will serve to achieve the objectives that have been proposed as a group. They will allow knowledge, action and, therefore, change, because "if there is no change there is no PAR" (Riva, 1993).

A practical case: how did the L'Alfàs Del Pi Project arise and develop?

The municipal government of L'Alfàs del Pi had identified some information gaps for its planning and decision-making process, so it was necessary to elaborate a broader frame of reference. For this reason, it was necessary to collect variables of specific interest, relevant information and discourses provided by the stakeholders of the municipality, which were neither developed nor available. For this, the municipal government signed an agreement with the University of Alicante for the research group AEDIFICATIO to design and develop the relevant tasks in this area.

To fulfil these objectives, the research team planned a workshop whose purpose was to know how the stakeholders interpret and configure the space and the experiences that take place in the municipality. Another objective of the workshop was to identify those factors that are problematic in urban conformations and that affect the daily lives of the residents of L'Alfàs del Pi. Lastly, solutions, including possible alternatives, to these problems were presented.

The workshop took place during a single day and was divided into two sessions: the first was composed of four activities (mental maps; population, tourism and leisure; healthy municipality; and, participatory activities for landscape assessment). The second session consisted in the preparation of a SWOT matrix, which had as a focus the municipality in general, without specific themes. Both sessions and their subcomponents were presented to the participants. Results and conclusions to follow will be derived from the second session.

The workshop was attended by 43 people, with a majority of male participants (59.5%), who were freely distributed around the room, among the four tables that the technical team had prepared. During the first session, all the attendees shared activities with the rest of the individuals who sat at their table. In the second, the participants were divided depending on their role.

Given the high influx of local political agents, it was decided to divide the participants into two groups. Group one was citizens and group two was comprised of political representatives. This *ad hoc* design allowed the research team to nurture the project since it was possible to see the connection or disconnection between citizens and those belonging to the political class. In addition, sharing allowed the different stakeholders to become aware of the proposals and needs of the other group, producing in this way a feedback system that will contribute to future decision-making processes.

What is a SWOT?

The objective of this activity is for participants to identify the strengths, weaknesses, opportunities and threats they perceive within the municipality. With this, a subjective diagnosis of L'Alfàs del Pi was obtained in the topics studied, with the aim of solidifying the foundations and lines of future work. According to its abbreviations, the SWOT matrix defines the internal and external aspects that favour or inhibit the proper functioning of a municipality. The matrix is broken down as follows: strengths, weaknesses, opportunities and threats.

- (S) Strengths are the characteristics of the municipality that allow it to be promoted and meet the goals set.
- (W) Weaknesses refers to the aspects of the municipality that, in one way or another, hinder its proper development.
- (O) Opportunities refers to events or characteristics external to the municipality that can be used in its favour to promote development.

- (T) Threats are external events that affect the municipality negatively, and that are sometimes uncontrollable by their authorities and inhabitants.

To work on the matrix the following steps are followed: (1) identify one of the thematic areas on which you want to work, for example, leisure and tourism; (2) make a list of opportunities; (3) make a list of threats; (4) make a list of strengths; (5) make a list of weaknesses; and (6) incorporate them into the matrix. To finish the development of the conceptual model of planning, a table of results is proposed to relate the objectives, strategies and policies.

Results

Common aspects

STRENGTHS AND OPPORTUNITIES

Interesting results emerged from the SWOT workshop. The municipal strengths that were highlighted by the two groups mostly referred to the natural resources available in the municipality of L'Alfàs del Pi, to its cultural offerings and to the coexistence of populations of different nationalities. The citizenship table considers it unlikely that cultural offerings would constitute a municipal strength since the available activities are directed to some but not all population groups. The table makes reference, in spite of valuing it positively, to the scarce offers directed to youth, to the elderly and to the non-foreign population.

The coexistence of people of different origins is another aspect that is valued by both tables, since it makes L'Alfàs del Pi a municipality with wide cultural and religious diversity. It is noteworthy that the table of the political class seems to have a more idealised image of the foreign population since no element of conflict or social tension is indicated. Not all data has proven to be positive however. The citizenship table also indicates that cultural diversity can be negatively associated at times with isolation, with the foreign population receiving more favourable political attention than locals and an overall questioning of the tourism model. In this sense, the residential model is confronted with the seasonal model.

On the other hand, natural resources in the areas of Playa del Albir, marine reserve and Serra Gelada Natural Park make up another group of aspects that are valued by both tables. In addition to the natural resources, the climate-related problematics that come along are also noted in the tables. Both groups who participated in the workshop raised the issue of natural resources and the problems these resources experience. The fact that this issue was raised so quickly by most participants of the workshop indicates the high value this issue carries among the Alfasin population.

Aspects that are seen as opportunities for future reference mainly have to do with communication within the municipality and with nearby municipalities. In particular, it emerged that tourist and economic dynamism within the municipality would improve the Mediterranean corridor. In this way, routes, travel times as well as new tourist activities would be optimised within the region. Additionally, it also emerged that the proximity of Benidorm is seen as an opportunity as a source of jobs for the Alfasin population since it counts on a commercial centre.

WEAKNESSES AND THREATS

As often occurs in the elaboration of SWOT matrices, the weaknesses that are present in the municipality are the most numerous and the most generalised. Both tables coincide in many aspects, which can be grouped in two main factors: communications and transportation, and coexistence. The first factor puts the focus on the national road N-332 as well as on public transportation. The N-332 is presented as an architectural barrier that in addition to dividing the municipality into two, makes it difficult to access one area from another due to traffic and problems of pedestrian access. Both groups agree that this national road is the main cause of the coastal area not being associated with L'Alfàs del Pi by the foreign population. Foreigners tend to believe that Albir Beach belongs to Altea. This misconception harms the image of the municipality. On the other hand, both the citizenship board and the political class believe public transportation to be deficient. Both groups refer mainly to the issue of buses, which have no route through all the entities of the municipality. There is no bus line offered that goes to the Specialty Center of La Nucía, which would be an important location to access.

The second issue has been labelled as coexistence: participants refer to problems of communication between the city council and citizens, as well as the isolation of population groups, such as elderly people. The citizens highlight the lack of measures for the integration and improvement of coexistence. This last coexistence factor also includes, more broadly, aspects related to the natural and urban environment, highlighting recycling, cleaning and pruning containers (in general garbage) as municipal weaknesses. The low degree of public awareness and lack of involvement in these matters also appears to be a concern.

Other topics emerged from the questions that the groups answered. The proximity of Benidorm and Altea, besides being seen as an opportunity, is also seen as a threat, especially for issues related to tourism. It should be noted that both tables highlight the community's discontent with the fact that the population outside the municipality associates the coastal area with Altea. Another aspect that is indicated as a threat to the municipality refers to public administration and bureaucracy. It is noteworthy that at the citizens' table, the focus was placed on political representatives and the municipal government, while at the table of the political class, the focus was placed on supra-municipal institutions. In this framework, difficulties encountered by the regulations to promote the primary sector and industry in the municipality were highlighted. The need to promote other economic activities in both sectors is widespread. Finally, climate change is seen as a future threat for L'Alfàs del Pi, as climatic changes could jeopardise the chance of the municipality to be chosen as a touristic destination.

Discrepancies

Both tables disagree on some aspects, which we will discuss in this section. Political representatives considered security a positive value in the municipality, while the public thought otherwise, considering it a weak point. For citizens, a greater presence of the municipal police is necessary, since in areas like Albir there is a lack of vigilance given the robberies that occur in this area. On the other hand, political representatives referred to L'Alfàs del Pi as an innovative city, while citizens saw a lack of innovation, mainly in terms of mobility.

Another aspect where the two groups disagreed is in reference to sports facilities: citizens are committed to making the infrastructure of the region profitable, while the political representatives talk about creating municipal infrastructures instead. Finally, both tables differ in their opinion regarding the AP-7 motorway, although as we will see they focus on two different aspects. Citizens made reference to the fact that the AP-7 is favourable to the municipality since it improves communication channels. The political representatives, however, felt that the AP-7 was a weakness, not necessarily because of its existence, but because of delays caused by traffic.

Main conclusions of the SWOT analysis in L'Alfàs del Pi

The two groups are quite in agreement on what the municipal reality is. The groups only disagreed on a few issues which, as the participants themselves pointed out, can be solved with a more stable and fluid communication between the political representatives and citizens. In this analysis we have seen the construction process of the SWOT matrix, although it is necessary to emphasise that the workshop dynamics prevented the two tables from developing strategies to maintain or modify those aspects that they had pointed out as relevant. This is an aspect that would be interesting to consider in the future, because although the work conducted allows us to diagnose how the collective imagination thinks of the municipality, it would be valuable to define how to act going forward as well.

However, the work done shows enough to indicate at least two key aspects in the daily life of the people residing in L'Alfàs del Pi. On the one hand, the Alfasin population highlights a major positive value in the natural resources available in the municipality. On the other hand, both groups remarkably notices three main issues: urban issues (specifically about the architectural barriers that divide the municipal area), challenges in coexistence (measures of social integration and questioning of the tourist model) and disagreements about mobility (communication with other municipalities, urban and interurban public transport and travel time). The municipality's dependence on tourism becomes apparent. Tourism represents a key sector of the Alfasin economic dynamic. From this revelation, a general need to contemplate other economic alternatives (primary sector and industry mainly) has emerged.

To conclude, it is worth mentioning that in this chapter we have focused mostly on a few of the more relevant aspects obtained during the session. Nevertheless, other areas of less intensity, such as education, pollution, the question of water, health, housing and relations with other municipalities have surfaced.

Discussion

It can be challenging to ask the community to be involved in solving problems. Some of this involvement includes, but is not limited to, asking for public opinion on topics such as urban development and its subsequent evaluation (Morgado, 2013). It has become apparent that, in order to attain informed results, the entire population of the territory should be under study. In this case, it's the population of the Mediterranean territories that should be under study, also according to the Europe, 2020 Strategy guidelines.

The different sections that make up this chapter have tried to shed light on citizen participation processes. It has been shown that when referring to citizen participation, it is advisable to ask a series of questions, since depending on whether they are answered in one way or another, different meanings of this concept are present. What is citizen participation and how should we understand it? Who produces the data? For what and for whom? Whom does the data speak about?

Having said that, we should no longer fall into the most common misconception, which is to identify citizen participation with a positivist conception. In other words, citizen participation is often incorrectly translated with participation originated from external data, that is, outside citizenry, the use of which is common in generalist surveys. The use of this technique without taking into account the PAR strategy, within the framework of a mainly quantitative methodology, suggests some doubts about its coherence with the concept of participation. The generalist survey is based on a directive technique in which people are asked questions based on concepts that are defined by the research group, based on their conceptual vision of reality or that of the person who finances it. Of course, this does not mean that the survey has no utility, or is not valid to capture a part of reality, but an even better survey would connect the notion of citizen participation with a concept closer to citizenship.

When discussing data production within the framework of citizenship, it has been proven that the use of more refined techniques is more effective both for the involvement of citizens and for the production of knowledge closer to social reality. In addition, since citizens define what is important and why, the survey generates an active feeling of belonging to the population in the decision-making processes. All efforts by local governments will report improvements in the municipalities. For example, a simple and accessible element for any city council are "citizen mailboxes", which can be incorporated both in the physical space and in the digital one, on the website. These mailboxes allow citizens to pose to the local government problems, suggestions or possible complaints of any kind.

When it comes to initiating decision-making processes that affect the whole of the community (economically, politically, culturally or socially), the rest of the qualitative techniques developed from the PAR are those that answer the questions raised in the investigations. Citizens themselves define what is important and endow their subjective meanings with the concepts that arise in the process. There is very little directive technique, where the research group focuses on the methodological aspects, rather than on what is produced. In other words, the research team is responsible for offering different means (techniques and tools) so that the groups that participate in the process can express and reach conclusions freely.

Thus, the nature of PAR as a social research process is at the same time a research methodology and a process of social intervention, as indicated by Basagoiti, Bru, and Lorenzana (2001). The analysis of reality is a form of knowledge and awareness of the population itself, which becomes both an active subject and protagonist of a project of development and transformation of its immediate environment and reality. In this way, the co-production of information, data and actions – complex and sometimes latent elements of reality – is achieved by researchers and actors involved in the day-to-day affairs of the subjects in question.

Note

1 Following the classification made by Morgado (2013), citizens can play a triple role in the development of social policies: the role of promotion or generation of policies, the role of intervention in programmes and the evaluative role of public and private policies. The degree of involvement of community members will determine the quality of the intervention.

References

Baño, R. (1998). *Participación Ciudadana. Elementos Conceptuales. En FLACSO, Nociones de una Ciudadanía que Crece*. Santiago de Chile: LOM Ediciones.

Basagoiti, M., Bru, P., & Lorenzana, C. (2001). *La IAP de bolsillo*. Madrid: Acsur-Las Segovias.

Canclini, N. G. (1995). *Consumidores y Ciudadanos*. México, DF: Grijalbo.

Cerritos, P., & Rodríguez, M. (2005). *Los Mecanismos de Participación y Concertación para el Desarrollo Local en el Salvador*. San Salvador: SACDEL.

Europe 2020 Strategy. (2010). *Europe 2020: The European Union strategy for growth and employment*. Retrieved from https://ec.europa.eu

Francés, F. J., Alaminos, A., Penalva, C., & Santacreu, O. A. (2015). *La investigación participativa: métodos y técnicas*. Cuenca: PYDLOS.

Gyarmati, G. (1987). La pedagogía de la participación: Una teoría política del bienestar psicosocial. In G. Gyarmati (Coord.), *Hacia una teoría del bienestar psicosocial* (pp. 231–244). Santiago de Chile: Pontificia Universidad Católica de Chile.

Ibáñez, J. (1985). *Del algoritmo al sujeto. Perspectivas de la investigación social*. Madrid: Siglo XXI.

Jiménez, B. (2008). *Subjetividad, participación e intervención comunitaria: una visión crítica desde América Latina*. Buenos Aires: Paidós.

Martínez, R. (2007). Desafíos estratégicos en la implementación de los programas sociales. In J. C. Cortázar (Ed.), *Entre el diseño y la evaluación* (pp. 63–118). New York, NY: Banco Interamericano del Desarrollo.

Montañés, M. (2009). *Metodología y Técnica participativa. Teoría y práctica de una estrategia de investigación participativa*. Barcelona: UOC.

Morgado, J. (2013). *Participación ciudadana y visiones sobre la política social* (Tesis doctoral). Universidad de Chile, Santiago de Chile.

Riva, F. (1993). Investigación Participativa y Autoformación Grupal. *Documentación Social, 92,* 141–152.

Riva, F. D. (1994). *Gestión participativa de las asociaciones. Segunda parte. Selección de lectura sobre trabajo comunitario.* (G. Bustillos, ed.). Managua: IPADE.

Seymour-Rolls, K., & Hughes, I. (2000). Participatory action research: Getting the job done. In I. Hughes (Ed.), *Action research electronic reader* (pp. 13–47). Sidney: University of Sidney.

Sojo, C. (2002). La Noción de Ciudadanía en el Debate Latinoamericano. *Revista de la CEPAL, 76,* 25–38.

Touraine, A. (1995). *¿Qué es la democracia?* México, DF: Fondo de Cultura Económica.

UNDP. (2004). *United Nations development programme.* Disponible en www.pnud.cl

Valero, A. (2005). Las perspectivas de la investigación social. Del pensamiento restringido al pensamiento complejo. *Saberes, 3,* Separata.

8 L'Alfàs del Pi

Mental maps

Pablo E. Vengoechea

Introduction

The perception of city spaces and the images inhabitants have of their surroundings are an invaluable tool for understanding built and unbuilt territory. This chapter focuses on the importance of the physical and emotional responses to urban space and habitat as a guide to better planning, urban design and a healthier city. The images and maps of the city generated during the public participation workshop discussed in Chapter 7, "Citizen Participation in Natural and Sociocultural Landscapes: Methodology and Main Results" represent memories and experiences that will be utilised to help interpret qualitative and quantitative information and guide the development of a plan, or targeted development projects, that improve the quality of life and health of the city.

As an architectural practitioner in New York City, I have used mental maps in community planning workshops to help inform how we might rethink and reorganise the urban spaces of the city. In the instance of L'Alfàs del Pi, the maps have served to identify what is most significant or problematic, culturally and physically, to the inhabitants regarding the city and helped prioritise future interventions. The drawings included here convey both existing information and the possibilities of change.

The methods used to acquire the sense citizens have of their city and the analysis which follows are rooted in the work of Kevin Lynch, urban planner and author of several publications including *The Image of the City* (Lynch, 1960). In it, he introduces the use of cognitive mapping as a way to understand residents' perception of the urban context and how they navigate through the city. Most importantly for us, it suggests how we may employ those perceptions and memories and incorporate, as outsiders, that intimate knowledge of the city and its form in the development of planning and development goals.

Cognitive maps and planning

Cognitive maps are drawings that represent how people acquire, classify, store, recall, evaluate and use information about the character of their spatial surroundings and their relationship to it. Such drawings contain approximate locations of people and objects in the natural and built environments and their essential attributes and are an important component of their spatial decision-making process.

This is the process that helps us navigate our environment based on our knowledge and perception of external conditions, the mental pictures we have of places and our attachment and attitude to such spaces and places. Built upon a combination of elements taken from several disciplines, drawing cognitive or mental maps provide a conceptual framework for exploring and interpreting perceptions of the city and designing appropriate place-based interventions for urban designers, landscape architects and planners, as well as government and administrative entities.

Planners and urban designers study this kind of mapping to understand how landscape and urban space affect people's interactions and identification with place, and by behavioural geographers who focus on the cognitive processes underlying spatial reasoning, spatial decision-making and behaviour. These geographers consider that what the individual knows or believes about the world helps to explain what the individual does or will do in order to make choices about the spaces and places that person inhabits. (Downs & Stea, 1973). Thus, the actions and mental states of individuals cause and are caused by perceptions of the social and physical environments, and are influenced by the spatial context, landscape patterns, group identity and ongoing interactions (Lynch, 1960; Relph, 1976).

Why we use mental maps

In my own planning work, professionally and academically, I have found that the experience and images that people have of their neighbourhood or city provide invaluable information on a community's sense of itself and the significance of the built environment. The images are important because they immediately disclose the issues that have bearing on the area's residents and quality of life. The information that can be gained from them in one-day events is on a par with traditional survey methods or specialised studies. The creation of mental maps in public workshops such as that carried out by AEDIFICATIO on behalf of L'Alfàs del Pi facilitates access to deeply rooted knowledge about the built environment and enables planners and participants to very quickly find common ground to resolve opposing positions, find shared solutions and arrive at a consensus on priorities.

Participatory planning processes and the use of mental maps reveal points of view and community information that may be missed or deemed unimportant, particularly by planners relying on Big Data to make decisions or needing the certitude that data might provide. The urban plan developed with this practice in mind is in effect more precise because it incorporates the shared perception of urban spaces and places in the analysis. The images created denote memories and experiences as well, which give a place an identity and can serve to guide the development of local neighbourhood planning and urban design projects. As we can discern from the one-day workshop held in L'Alfàs del Pi, participants not only described their movements through the city and identified important external reference points but also communicated their personal experiences with the city's public streets and gave them meaning through memories of events and the familiarity associated with both desirable and undesirable areas of the city.

Another reason for the importance of mental maps is less quantifiable but has everything to do with the idea of place-making and identity. An identity that is local and built on a foundation of site, place and the iconographic value and attachment to landscape and tradition, which today is being rapidly eroded as cities replicate generic versions of tired urban renewal solutions and embrace universal solutions without regard for the genius of a place (Norberg-Schulz, 1980).

Thus, the plan the city of L'Alfàs del Pi intends to develop should incorporate the territorial knowledge of the city, the socio-cultural and political polemics and the concerns of not only the native-born citizens but of those others that have rightly claimed the city as their own and self-identified as Alfasinos. Moreover, the plan for a healthy city should make use of this knowledge in shaping new urban space giving it meaning that reflects the consensus of the workshop participants.

Mental map exercise

As described in detail in the previous chapter, a one-day workshop with six activities took place in L'Alfàs del Pi on June 3, 2017. Over 75 people participated in this event or attended part of the day, which included residents, visitors, government officials and administrators and professors and students. Key among the activities held that day was the mental mapping exercise. Participants were asked to draw a personal representation of the city tracing either their movements and daily or weekly activities, and the places that held special meanings and memories for them. Then they were asked to share their knowledge, sense and understanding of their community with others in an open forum.

The images that were drawn are unique representations of community and city and, per Kevin Lynch (1960), contain five unique elements of identity, city form, ambiance, social structure and meaning, which can be analysed. They serve to orient and personify and give meaning to the spaces of the city. The five elements used to form mental maps are:

- Pathways and movement channels between places and activities: the streets, sidewalks, trails, canals, bikeways and pedestrian walkways, transport or vehicular channels in which people travel around the city, village, neighbourhood or site.
- Edges and perceived boundaries, real or imaginary: walls, buildings, changes in land use, typology, activities, terrain, built form, economic condition, culture and neighbourhood character, streets and highways, rail lines, parks and shorelines.
- Districts and neighbourhoods: medium and large, cohesive areas of the city distinguished by a concentration of shared physical, social, economic, cultural characteristics or defining identity and clear boundaries and entrances.
- Nodes, focal points, intersections or loci: points, places or strategic spots that have a concentration of activity or are a focus of the community and serve as focal points. Such places might include streets, busy intersections, shopping areas, town centres and boardwalks.

• Landmarks: identifiable objects that aid in orientation and way-finding for neighbourhood, city and region. Such places are external references and can be as large as a mountain or the seashore, or smaller as a school, house of worship, lighthouse, antenna or advertising sign or public art.

Instructions to workshop participants

After a brief introduction on mental maps and their importance to the planning and design process we were undertaking, each participant was tasked with drawing her/his impressions of L'Alfàs del Pi. Participants were asked to respond to several questions and illustrate their responses by creating personalised maps, and/or pictures and symbols of the city using the five fundamental spatial elements described above.

The questions included the following: What do you consider the boundaries of the neighbourhoods and the city to be? What are the most important places in your community, and in the city? What are your most travelled paths? What are the natural or man-made edges of the L'Alfàs? What are the most active areas and why? Which places have the most personal value? Which places are the most memorable areas of the city? Which are the least memorable? Why? What are the community's most valuable assets and irreplaceable resources – natural and built?

Our objective was to generate discussion and rough illustrations as instruments of information and communication. Each of the participants drew a sketch map of the city according to their own viewpoints with the goal of identifying how urban space is used and their spatial sense of the city.

Analysis

The following images and mental maps were drawn by the residents of L'Alfàs del Pi and speak to the importance of certain spatial elements and the character of its built and unbuilt environment. The sketches offer clear images of what is most important to the residents, which, depending on the individual, could be simple or complex and directed toward specific concerns. The sketches represent fragments of a shared reality that, although drawn from memory and not quite geographically accurate, unveil the essence of a city.

We looked at these mental maps for the personalised information that encapsulates their sense of the city. In my practice, we recognise, as architects and urban planners, that the act of drawing is a tool not only for communicating ideas and visual representation but also for thinking and processing information. With my students and in the community planning workshops I have conducted in the past, some of the most revealing moments come when a participant's resistance to drawing a line out of fear that it is not a good representation of reality is overcome and the individual becomes aware that the line, scribble or symbol that has been drawn has the power to represent a fact, an idea, a pathway, a process, an emotion or a place. At that moment, the images created are more confident and become powerful tools that connect citizens as they identify where and what is happening in their neighbourhood, their shared concerns and their sense of belonging. As themes begin to emerge we begin to achieve consensus on the possible directions a plan may take and together weigh possible recommendations.

The graphic responses have been examined and tabulated to extract a series of conclusions regarding the ambience and needs of the city, and the practical use of the territory on which general and specific recommendations can be based. This analysis served to identify the unique components of the city's urban structure and will help the city to prioritise areas of improvement and develop a plan for a healthy and innovative community.

Below is a summary of the most frequent themes among participant's mental maps for each respective category – pathways, edges, districts, nodes and landmarks:

Pathways and movement channels between places and activities   Not surprisingly, mention of N-332 by far exceeded all other pathways in importance, followed by the streets crossing or leading to it, Cami del Mar and Av. de L'Albir. The Camino del Faro, a hiking trail, was also ranked very high. The Autopista de la Mediterránea, the rotunda and local neighbourhood streets also received notice. Interestingly, the railroad and the Passeig de les Estreles received an equal number of mentions, although the Passeig was also noted as a place of recreation, eating and shopping (see Figure 8.1).

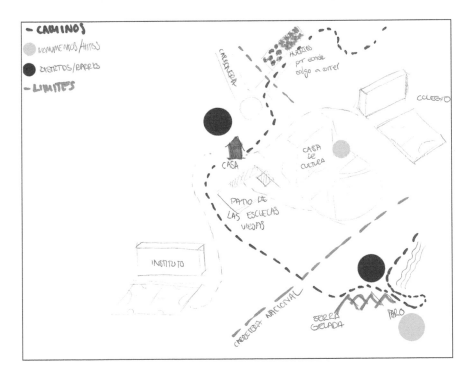

Figure 8.1 Pathways and movement channels.

Edges and perceived boundaries, real or imaginary N-332 once again was noted as the most significant boundary and edge condition, dividing the city such that L'Albir seems to become its own centre with different land use, built form, economic condition and culture. The steep topography of the natural hillsides and parks and the agricultural land's patterns and networks were also identified as important edges that define the city (see Figure 8.2).

Districts and neighbourhoods The most important district illustrated by the majority of participants was L'Alfàs del Pi's older precincts and the historic civic centre. L'Albir was the second most frequent neighbourhood illustrated in the minds of the participants, followed closely by smaller residential neighbourhoods (see Figure 8.3).

Nodes, focal points, intersections or loci L'Alfàs' natural features and climate, the Serra Gelada, the sea and the green zones composed of parks, citrus orchards and pine groves, and slopes constituted by far the most significant element and focus for people in the workshop. The historic core of the city and its civic centre followed closely behind as did the beach and restaurants along the Passeig de les Estreles much further down. The Casa de la Cultura was also identified as an important node and strategic spot. Lastly, several nearby municipalities (Benidorm, Alta, La Nucia and Alicante) were mentioned as centres of activity (see Figure 8.4).

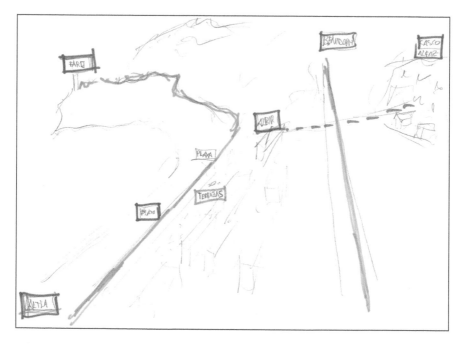

Figure 8.2 Edges and perceived boundaries.

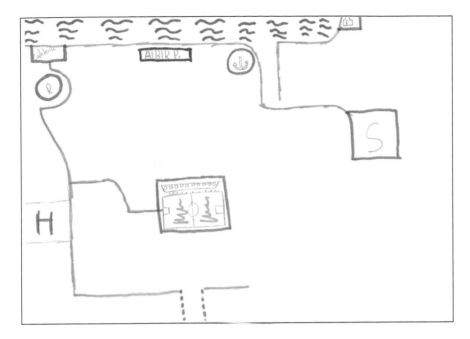

Figure 8.3 Districts and L'Albir neighborhood.

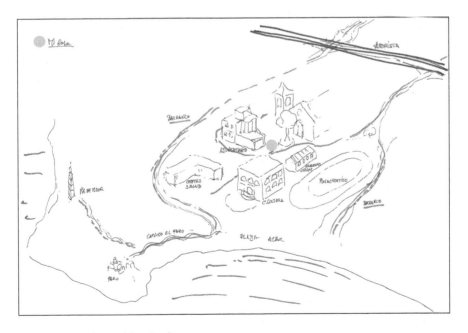

Figure 8.4 Nodes and focal points.

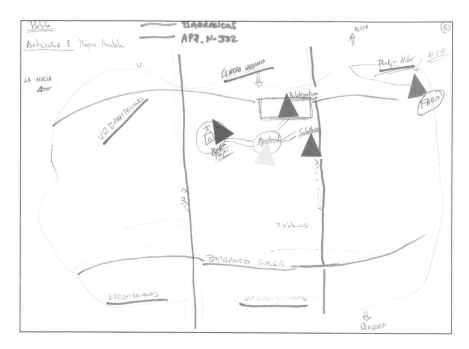

Figure 8.5 Landmarks.

Landmarks   The number of landmarks listed as identifiable external references for orientation and meaningful symbols of L'Alfàs included the Faro del L'Albir, followed by the tree that is the symbol of the city, the Polideportivo and the Church of St. Josephs. This list also included supermarkets, municipal services (police, health centres and a school), the Fundación Frax and archaeological ruins (see Figure 8.5).

General findings, themes and emerging topics

The workshop exercises gave us vital information on the environment and provided us with a font of factual as well as perceptive information regarding the elements and issues most important to residents. In general, we can detect several tendencies and conditions that can be used to improve the city for all, prioritise areas of planning interest, rethink patterns and establish a vision for the city's future.

Public health   Many individuals included references to the beneficial climate, fresh air and the abundant natural areas of L'Alfàs del Pi. The latter figures prominently in more than a few of the drawings containing commentary and images denoting a brilliant sun, the Mediterranean Sea and the beach, the Serra Gelada and the hillsides, along with the fauna and avian life, land uses and sporting and social events associated with these places. Appreciation for the beauty of its location should be incorporated in any planning for the city and future projects; specifically, natural areas should be protected and a tree planting programme initiated (see Figure 8.6).

Symbols of the city   These include the historic tree at its founding core, the old city and the government centre and the Faro de L'Albir, the Casa de Cultura and the Fundación Frax. With these emblematic cultural places it would be possible to explore and construct a new narrative for the future as a healthy and innovative territory, grounded in the history of L'Alfàs del Pi and its current multicultural and multiethnic character. The identity of a place has three components: physical setting and natural landscape, which act as backdrops for activities, which are influenced by the backdrop and a set of meanings associated with those activities (Relph, 1976). If this meaning is rooted in the setting, objects and activities, but not a property of them, it is clear that L'Alfàs del Pi has the people and the necessary components for creating a new pluralistic narrative (see Figure 8.7).

Carretera Nacional 332   One of the first and perhaps most important areas of interest identified by nearly all participants in the workshop concerns the N-332. With few exceptions, the city appears to the majority of participants to be physically and socially divided along the axis of the north-south road. This perception is supported by the frequency the roadway appears on the mental maps and the manner in which it is illustrated, and confirmed by our analysis (see Figure 8.8).

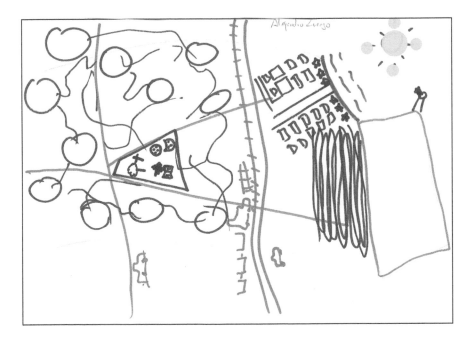

Figure 8.6 Climate and nature.

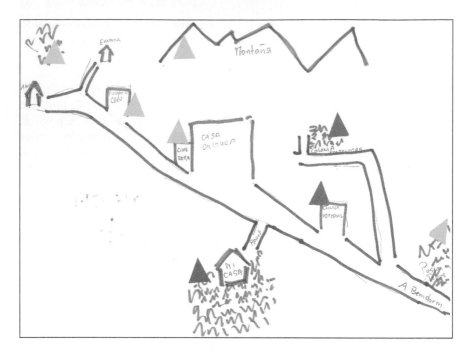

Figure 8.7 Symbols of the city. The Casa de Cultura and the mountains.

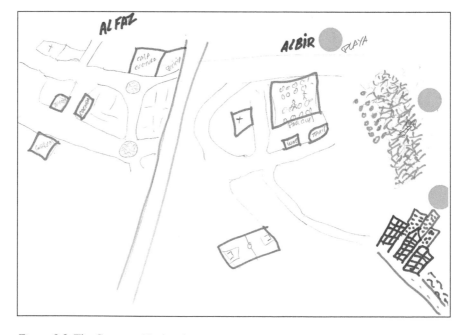

Figure 8.8 The Carretera Nacional.

The meaning of the road is what we must consider at this stage, and the opportunity that it represents to resolve other structural community and planning issues, as well as the identity of L'Alfàs del Pi. The N-332 is often the first area of the city one experiences, and is not a very welcoming sight. In discussing and presenting the maps many pointed to the negative quality of activities and land uses found on the road, and to the traffic conditions. The limited crossings and the existing rotunda are problematic. This situation was felt to further divide the city and several individuals expressed interest in integrating the two areas of the city through culture and education. The breach itself can be overcome by, at a minimum, creating new crossings and improving the existing crossing, and at best reinventing this stretch of N-332 as a destination street, creating a place not to bypass but to stop and meet people, even as it fulfils its transportation function. It is an opportunity to create a distinct, shared urban space and place that is contemporary, attractive and has a visual appeal.

Two cities, L'Alfàs del Pi and L'Albir Tellingly most participants focused on one or another part of the city in their habitual routines and emphasised their travels at either side of route N-332. Several participants focused almost entirely on L'Albir, the residential area by the sea, east of the N-332. Most activities (cultural, social, recreational, business and shopping) for these individuals took place within this area and most perceived their neighbourhood to end at N-332. These individuals appeared to generally have less connection with the historic core or a need to move beyond these boundaries. For others, the focus of their activities was L'Alfàs del Pi or more specifically the area west of the road, generally because it was their place of residence, work or recreation. In those instances there appeared to be an appreciation for the local services, the seat of government and the places of memorable events and the significantly older, tightly knit urban fabric and architecture.

While this east-west separation was evident in several cognitive maps, food places such as restaurants or supermarkets easily brought people across the road. These two areas have distinct demographic characteristics and are very different territories. One almost has the sense that L'Albir is its own city in some of the maps, a situation that should be addressed in the future (see Figures 8.9 and 8.10).

Cultural diversity The mental maps also noted differences in demographic and cultural characteristics. A desire for places where the various communities can come together and share histories and traditions, emotional energy and a visible commitment to the city and the natural surroundings that all enjoy was evident in several of the maps and discussions, as well as an expressed desire to meet the needs of the various constituencies.

Still, for others, the idea of home was very important and the path to their homes, perhaps as a place of refuge and safety and a place from which you begin to recharge or to understand and confront the world. In sum, there were maps with non-place-based elements that acknowledged and spoke to the idea of coexistence, and captured the essence of a pluralistic society (see Figure 8.11).

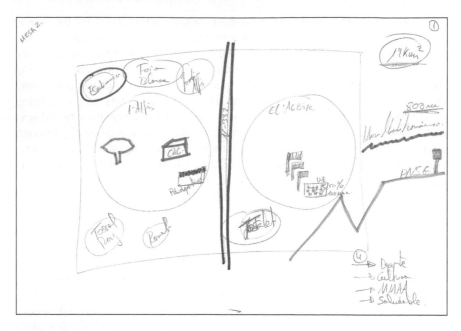

Figure 8.9 Two cities.

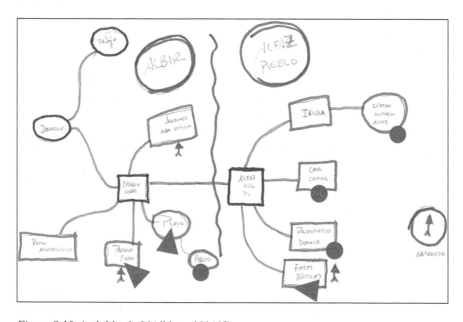

Figure 8.10 Activities in L'Albir and L'Alfàs.

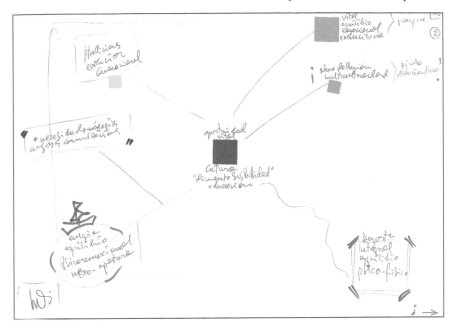

Figure 8.11 Cultural and demographic diversity.

Region   Lastly, for some individuals the regional context and situation of the city in relation to other municipalities, agricultural lands and natural features would seem to indicate a yearning for a deeper connection with landscape and people (see Figure 8.12).

With this information, it is possible to for us to venture into building the narrative for the future of the city as a healthy and innovative territory. It's important at this stage to reiterate the sense of belonging and connectedness citizens have with the city of L'Alfàs del Pi and the natural landscape that surrounds it. The growth of the city should be directed away from sensitive natural areas in order to protect its agricultural heritage, open space and scenic landscapes. It is the seed of a vision and the beginning of a formal plan and decision-making process in which N-332 will play a major role.

Next steps: an urban design plan for N-332

This section will address preliminary urban design considerations necessary to transform this stretch of a roadway into a great destination street with unifying urban elements and notable legibility. In a sense, the roadway has multiple meanings – it is simultaneously a through-way and a gateway, a destination and a place to bypass and a barrier and a means of connection. It is a place with an industrial

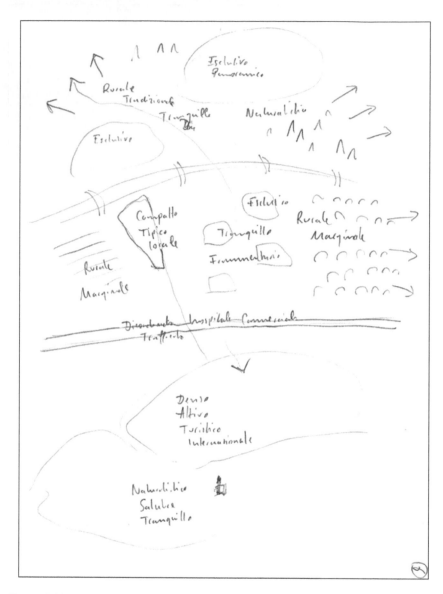

Figure 8.12 Natural context.

past and unwanted uses and activities, and a well-used and congested road that currently divides the city into distinct areas with different socio-economic and cultural characteristics. The road, which has been described with various epithets – inhospitable, congested, ugly, unwanted, undesirable, barrier – is in serious need of attention. It divides the city and there aren't enough places to cross it, to the extent that L'Albir begins to be perceived by many as separate and not a part of L'Alfàs del Pi.

At this early stage, we would seek to understand for whom the project is being planned and elicit participation – listening and learning from citizens, defining the purposes for the transformation of the corridor and identifying a broad vision. In essence, we need to get to know and document the place and fully understand the social, cultural, political and economic contexts. Such a site analysis should include sketches, measurements, mapping data and public input.

Once all voices are heard and understood, and different parties are involved, an urban design project coordinating the spatial and social complexities of the roadway can be initiated. We believe this space can be transformed from a heterotopic and pathological urban condition (Foucault, 1967) to a great boulevard with qualities that give meaning to the roadway as a destination for work, recreation and culture. This section of roadway could become a primary and vibrant element of the city contributing to its economic growth, urbanisation and development.

Designing such a place is complex; it should be a participatory process and incorporate ideas across different areas of expertise, interests and scales in order to ensure diversity and sustainability. A people-centered and sustainable urban design approach will take into consideration key issues in the community with quality public places, supporting services and the right land uses and a mix of activities. This participatory process will also help to identify a wider range of strategies and possible investment opportunities that support the city's growth, needs and cultural vitality.

A core value of this work would be to ensure the implementation, livability and maintenance of this new public realm. The aim is to create a great corridor, one that creates a place, has a clear identity, is familiar and visually engaging and has meaning for the residents even as it continues to serve a regional transportation and infrastructure function and is accessible to all. It should also be cost-effective: the investment made by the governing body on these improvements should be offset by the return. These costs can be measured at the different scales at which investments can be found to exist – citywide, neighbourhood and local.

The roadway corridor is an opportunity to celebrate L'Alfàs del Pi's history, its people and the culture of those that call the city home. The urban design plan should create a rich and welcoming street that is an iconic destination with vibrant public spaces. It is time to reclaim the N-332 roadway and bring it back into the fold of a unified city.

On making a great urban corridor

A great street is superior in character and quality and a place that people visit often, it contributes to the community, it is comfortable, safe and encourages social interaction, it is memorable with a distinct identity and representative of the city and its residents (Jacobs, 1993). Such great streets contribute to the community and reflect the local culture and context. There are large- and small-scaled great streets, narrow and wide, some are arterials, others are local streets, all have a strong sense of place, are attractive, provide safe spaces for the pedestrian, are economically vibrant and diverse, balance different transportation modes and are responsive to climate. Such streets include La Rambla, Barcelona, a tree-lined pedestrian thoroughfare stretch always crowded with people, and Bourbon Street in New Orleans.

Streets with comparable transit and high traffic counts include Champs Élysées in Paris, Istiklal Street in Istanbul, which although much narrower than N-332 has several streetcar lines and it is a busy shopping street, and Avenida 9 de Julio, Buenos Aires, perhaps the widest street in the world. It accommodates nine lanes of traffic in each direction, wide sidewalks and landscaped pedestrian areas, underground parking garages and public transit comprised of metro stops, trams and most of the city's bus lines. The street is also punctuated by historic civic monuments, which add a distinct dimension to this very busy thoroughfare.

I recognise that this is not La Rambla in Barcelona or Fifth Avenue in New York City; nevertheless, the road can and should be transformed into a lively transit corridor with people-centred public spaces. As shown in research conducted by William H. Whyte published in *The Social Life of Small Urban Spaces* (Whyte, 1980) "what attracts people most, it would appear is other people". The impact of that work and others regarding the nature of public spaces and the recipe for creating successful public spaces have become part of the toolkit and instruction manuals used by urban designers and planners worldwide.

I do not speak of an opportunity to functionally improve the street in terms of the necessary regional and local services that it provides, such as efficient traffic movement, though this will also be addressed in its redesign, but in terms of its condition as a place that serves a social purpose and has unique community building aspects. It is to the latter that these initial thoughts are noted, with the understanding that the technical traffic, circulation and functional requirements of the street will be amply addressed and improved upon by transportation planners and civil engineers during the reconstruction that sooner or later must take place. What I consider equally important and urgent, and what may not receive the attention it deserves, is the reinvention of the corridor as a vital public place. A place that unifies rather than divides, that is a destination and a place of interaction, reimagined and reconstructed with design characteristics and land uses intended to promote the development and celebration of community and public life.

The reimagining of this corridor must also include adjustments to the existing ordinance, which I am guessing is regulated by a zoning code that is probably obsolete, put in place for the industrial uses of another era. Outdated regulations may be partially responsible for unproductive roadside activities and do not contribute to the quality of life or health of this community. These activities and the former industrial uses still found on this road are no longer viable; these industrial spaces and areas are drosscapes (Berger, 2006) that have been taken over by a host of unwanted uses. Areas like this are typically found on the periphery of the city and are the byproduct of rapid urbanisation and growth that leapfrogged the old industrial sites and outmoded industrial production systems and technology, which are no longer economically viable. These wastelands result from the process of deindustrialisation and should and could be recovered, reprogrammed and reintegrated into the fabric of the city. The city must address this major infrastructural design challenge and take on the job of reimagining and reclaiming this urban corridor.

L'Alfàs del Pi, like many cities built in another time, has accommodated the presence of the automobile by decreasing pedestrian space available; often sidewalks are removed or narrowed as the roadbed was widened. While the need to accommodate the car and modern vehicular traffic is important to a contemporary city, the fact remains that it has increased the distances between neighbourhoods and between the core and the outlying districts of the city. It has made possible the separation of communities that are primarily connected by road networks, a model of development that will no longer be sustainable in the near future.

A development programme for improving N-332 must address the physical qualities of the corridor and the level of comfort users have when in such a place. This level of comfort is as much about design as it is about the activities that can be found in the place. A comfortable place is a place where one would want to spend time. The development programme should address the legibility of the street and proposed buildings, its urban character as defined by general activities and ground floor use and special streetscape improvements with sustainable urban features and a tree planting programme. Such a strategy is intended to reduce the carbon footprint of the street and improve its microclimate, and promote a human, not automotive, scale and speed.

Just as important will be creating a lucid narrative for the new urban corridor. What is its storyline? What is its purpose? What are our hopes and aspirations? What is the street's relationship to the surrounding physical and social contexts? A great street is a place for walking, a place that is well defined by its buildings and urban scape, is easily identifiable as a destination and place for civic interaction, has a vibrant visual character and coherence that catches the eye and has multiple uses and social purposes. We support the creation of a place that brings people together and encourages socialisation and citizen involvement in building community. Streets are the public realm of the city; this section of the N-332 can and should be reinvented as an urban corridor that contributes to the quality of life and economic engine of L'Alfàs del Pi.

Urban design considerations

This type of study is neither tidy nor sequential, yet it yields valuable results, which must be communicated to civic and professional organisations and non-government groups, architects, planners and businesses and citizens and builders. It has long been known that pedestrian pathways are important components of public open space networks in any city (Gehl Institute, 2018). The new urban corridor has to work for people and its public spaces should be interconnected, interesting and inviting, encouraging people to walk and be physically active. These spaces can go a long way to help L'Alfàs combat obesity, diabetes and stress and promote a healthy lifestyle consistent with the goals of the city (AARP & Project for Public Spaces, 2008). Such considerations need not hinder the efficient flow of vehicular traffic.

It's always of great value to remember the pedestrian is the active participant of this story; the improvements should be designed with that goal in mind. The corridor that is envisioned will transform this obsolete and forsaken stretch of road, become an economic engine for the city and improve the quality of life for all Alfasinos. Such a makeover can be combined with decisions about density, land use, multipurpose and sustainable infrastructure, traffic calming strategies, walking distances to residential areas, scale, public transit stops, lighting, parks and of course commercial and cultural amenities and programming.

We advocate more people-meeting-people spaces, communicating and exchanging information. We advocate places that feel secure for all and are welcoming, with well-defined public open spaces. We advocate compact urban development, walkability, infill design and community-oriented streets and neighbourhoods. We also want to reintroduce ecological networks in the structure of the corridor and provide people with daily contact with nature, which was very important to Alfasinos.

Climate change considerations

We live in an age where it is now the urgent responsibility of planners and governments to boldly address climate change in every aspect of city growth and management (*United Nations sustainable development goals: 17 goals to transform our world*, 2018). It's no longer sufficient to advocate for beautiful, well-designed, people-oriented public spaces and safe transportation – we need to become design advocates for sustainability. Hence, the urban design strategy should incorporate ideas and features for a new roadway and urban corridor that addresses climate change and recognises that resources are finite through the use of sustainable growth and urban heat management strategies, green infrastructure and vegetation and cradle-to-grave materials. This could include incorporating new techniques, such as the production of concrete that stores CO_2 and helps reduce a project's carbon footprint.

Climate change is a condition that is affecting cities everywhere, disrupting weather patterns and causing more extreme weather events. Surface temperatures continue to rise alongside greenhouse emissions and are projected to surpass existing average temperatures by several degrees within a few years (Wheeler & Beatley, 2014). To address climate change, most countries signed on to the Paris Agreement on December 2015. In the agreement, all countries agreed to work to limit the rise of global temperature.

Climate change, excessive use and reliance of motor vehicles for daily movement, unchecked growth and sprawl, reduction in open space and habitat areas, pollution, profligate use of natural resources, rising social inequities, loss of indigenous landscapes and ecosystems and a global economic system that undercuts local building and artisanal traditions and local homegrown commerce, are all areas of concern that can be addressed within the redesign of this new corridor infrastructure. The development of this piece of infrastructure, if properly man-aged, can be used to reduce the city's urban heat island effect (the rise in tempera-ture in built environments due to the lack of vegetation) and lower greenhouse emissions, and help manage air quality to improve the living conditions of its residents.

Conclusions and a call to action

Addressing these developments is no small task for governments and urban cen-tres such as L'Alfàs del Pi and by extension other Mediterranean cities. Urban centres need to very quickly get on board to find solutions to reduce unsustainable development practices and find ways in which growth can help the environment. The challenge will be reaching cities most in need and population groups that suf-fer the greatest health problems.

The roadway will never be a fully pedestrian street, nor do we want it to be that. At a minimum, it could be redesigned as an enhanced transit corridor to improve transit and pedestrian opportunities for residents, businesses, culture and recreation; or ideally, it could be designed as a piece of green infrastructure and landscaped transit corridor with improved local service connections.

Historical places such as L'Alfàs del Pi contain valuable experiential lessons, consequently in planning for the growth of the city; the new road can be singular (imbued with a sense of place) and universal at the same time (familiar, comfort-able). Singularity speaks to the uniqueness of that space, both morphological and social. The character of a place has a direct relation to genius loci, which denotes the essence of place (Norberg-Schulz, 1980). It is an old concept, the belief that all geographical sites and historic places are unique and have a guardian spirit, which gives life to people and place. Connection with this spirit is integral to the reinvention of the Carretera Nacional 332 and should be part of its identity.

> The identity of a place is comprised of three interrelated components, each irreducible to the other – physical features or appearance, observable activi-ties and functions, and meanings or symbols.
>
> (Relph, 1976)

References

American Association of Retired Persons and Project for Public Spaces, Inc. (2008). *Streets as places*. New York, NY: AARP. Retrieved from www.pps.org/product/streets-as-places-using-streets-to-rebuild-communities

Berger, A. (2006). *Drosscape: Wasting land in urban America*. New York, NY: Princeton Architectural Press.

Downs, R. M., & Stea, D. (Eds.). (1973). *Image and environment: Cognitive mapping and spatial behavior*. Chicago: Aldine.

Foucault, M. (1984). Of other spaces, heterotopias. *Architecture/Mouvement/Continuité*. Lecture given by Michel Foucault, March 1967. Retrieved from http://web.mit.edu/allanmc/www/foucault1.pdf

Gehl Institute. (2018). *Inclusive healthy places*. Retrieved from https://gehlinstitute.org/work/inclusive-healthy-places/

Jacobs, A. B. (1993). *Great streets*. Cambridge, MA: The MIT Press.

Lynch, K. (1960). *The image of the city*. Cambridge, MA: The MIT Press.

Norberg-Schulz, C. (1980). *Genius loci, towards a phenomenology of architecture*. New York, NY: Rizzoli.

Relph, E. (1976). *Place and placelessness*. London: Pion.

United Nations sustainable development goals: 17 goals to transform our world. (2018). Retrieved from www.un.org/sustainabledevelopment/sustainable-development-goals/

Wheeler, S. M., & Beatley, T. (Ed.). (2014). *The sustainable urban development reader* (3rd Ed.). London and New York, NY: Routledge.

Whyte, W. H. (1980). *The social life of small urban spaces*. New York, NY: Project for Public Spaces.

9 Urban and interurban mobility, green corridor definitions and strategies of intervention

Armando Ortuño Padilla, Jairo Casares Blanco

Territorial structure and current situation of mobility in the cities of the Spanish Mediterranean

Territorial structure of medium-sized cities of the Spanish Mediterranean and problems of mobility

Over the last few decades, the territorial model of the municipalities of the Spanish Mediterranean has progressed towards a structure mainly characterised by low-density urban developments. The Spanish Mediterranean territorial model has also moved towards a phenomenon known in urban planning as "urban sprawl", which is characterised by having detached houses in residential areas and the segregation of land use.

This chapter will discuss a specific case of a medium-sized city by the Mediterranean coast that exemplifies the problems exposed: L'Alfàs del Pi. This municipality is located in the north of the province of Alicante and possesses an enormous tourist potential that attracts thousands of foreigners – mainly Europeans – who have decided to settle permanently in the area or reside for long periods of the year (see Figure 9.1).

Territorial structure analysis and its problems of mobility associated in L'Alfàs del Pi (Alicante, Spain)

Analysis of the territorial structure

Placed in Benidorm's proximity, one of the main tourist attractions of Spain and Europe, L'Alfàs del Pi is shaped by a total of 12 consolidated urban areas resulting from the implementation of low-density urban development. The total population is 21,494 inhabitants (Instituto Nacional de Estadística, 2016). Three population centres emerge inside L'Alfàs del Pi's municipality: L'Alfàs del Pi's town centre, L'Albir and the development concerning the beach of L'Albir (see Figure 9.2). These three population centres make up two-thirds of the total population of L'Alfàs del Pi. The other third of the population resides in the other nine

Figure 9.1 Location of L'Alfàs del Pi (Alicante, Spain).

Source: Prepared by the authors.

consolidated areas developed in other locations, as well as other buildings spread in rural areas.

In addition, the municipality is divided in three parts which are separated as a result of a significant "barrier effect" generated by the intersection between highway AP-7 and the national road N-332. These two roads interpose themselves between L'Alfàs del Pi's town centre, the area of L'Albir, Serra Gelada and L'Albir beach. Furthermore, the TRAM railway line runs parallel to the N-332, thus accentuating this barrier effect.

Finally, it is necessary to emphasise that the municipality is located between two important tourist attractions. The first attraction is the zone southwest of Benidorm's municipality. This zone represents one of the most important tourist destinations of Spain and of the Mediterranean; it is the Spanish symbol for the country's "sun and beach" type of tourism. The second attraction is located in the north and it borders with the municipality of Altea. Although with a minor

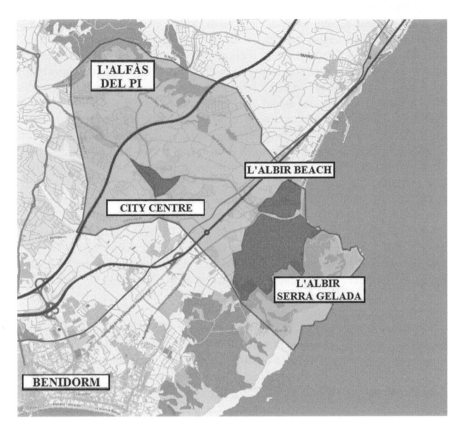

Figure 9.2 Major centres present in the city of L'Alfàs del Pi (Alicante, Spain).

Source: Prepared by the authors with data from the OpenStreetMap (2018).

relevance compared to Benidorm, Altea is also is an important tourist destination of the Costa Blanca.

Urban mobility analysis

As seen often in regions where the phenomenon of "urban sprawl" is observed, private cars are the most popular mode of transportation inside L'Alfàs del Pi's municipality. According to the *Study of Traffic and Mobility* of the city's General Plan of Urban Arrangement, the majority of transportation is conducted by car, with a 66.7% percentage rate (see Table 9.1). This last figure contrasts significantly with the much lower percentage of trips on foot, which is 22%. The percentage of public transportation being utilised is even lower, at 4.4% (Ayuntamiento de L'Alfàs del Pi, 2015).

It is interesting to stress that the heavy reliance on private car usage isn't similar in all population centres. L'Albir's Serra Gelada centre and L'Albir's beach are situated nearby centres where low-density is especially developed. Because of the low-density factor, these two areas display almost 60% of internal displacement and transportation between both centres is conducted mostly by motorised vehicles. L'Alfàs Del Pi's town centre, on the other hand, represents a more compact development model. The town centre is composed of a mixture of commercial businesses and tertiary use. Because of this more compact development, 70% of the trips inside the centre are made on foot. Certainly, these results already encourage gravitation towards a sustainable mobility territorial model. One of the objectives of this chapter will be to promote this model of urban development.

On the other hand, 4 km away, in the zone located between the population centres of L'Albir and the town centre of L'Alfàs del Pi, 62% of trips are conducted by private car. This percentage increases by 80% in the case of trips between the centres of L'Albir's Serra Gelada and the town centre of L'Alfàs del Pi. Although this distance may be excessive for walking, it would be feasible by bicycle; however, as can be seen from Table 9.1, the use of bicycles is also low, indicating a lack of adequate infrastructure, which keeps it from becoming a real alternative to private car commuting.

Finally, it is necessary to indicate that 80% of the total displacements happen by private car in the remaining urbanisations distributed within the territory of the municipality. Consequently, both public transportation and bicycling appear as residual models in the town, with, respectively, 4.4% and 1% of the total trips, as noted previously in Table 9.1.

Regional mobility analysis

L'Alfàs del Pi is placed between two important tourist areas. One of these areas is the relevant city of Benidorm. This strategic placement affects significantly the mobility of people within L'Alfàs del Pi. For this reason, travel from or to other towns represents 44% of the total trips from L'Alfàs del Pi. The principal town people travel to is Benidorm (50.1%), followed by the city of Altea (18.7%).

However, the importance of public transport for these types of trips is non-significant. It is necessary to emphasise the existence of the railway transportation

Table 9.1 Modal split in the city of L'Alfàs del Pi

Trips by modes of transport – percentages (%)							
Walking	*Car*	*Motorcycle*	*Truck*	*Bus*	*Railway*	*Cycle*	*Other*
22.0%	66.7%	4.7%	1.1%	4.1%	0.3%	1.0%	0.1%

Source: Prepared by the authors with data from the L'Alfàs del Pi city hall (2015).

system in the metropolitan area of Alicante, called TRAM. TRAM nowadays offers a service of one train per hour. The train operates with a displacement between 7 and 9 minutes up to Benidorm, between 5 and 7 minutes up to Altea and between 27 and 29 minutes up to Calpe (Ferrocarrils de la Generalitat Valenciana, 2018). Currently, the modal quota of the TRAM is less than 1% of the total trips. This fact is especially significant taking into account that the town offers two TRAM stops, L'Albir and L'Alfàs del Pi, which demonstrates that these services are available and yet enormously underused.

In the case of bus transportation, there are three regular intercity lines between Benidorm and Altea that cross the municipality, although two of them only pass L'Albir's centres, so they don't connect with other inner zones. In addition, there is an issue with bus stop accessibility around L'Albir's Serra Gelada centre. Because the maps for the bus stops are drawn without considering the residents living closest to Serra Gelada. One of these lines connects with three principal population centres of L'Alfàs del Pi (the town centre of L'Alfàs del Pi, L'Albir and L'Albir's beach). Nevertheless, this service of intermunicipal transportation is clearly insufficient for serving the area of influence since the stops leave out an important part of the population. For example, in the northwest zone, where 20% of the population resides, there is no regular public transportation service offered that connects the northwest with the rest of the municipality or with other nearby municipalities.

Definition of current problems in the area of mobility

Previous results allow us to determine roughly the main problems of L'Alfàs del Pi in terms of mobility. However, as mentioned previously, this town has a series of problems very similar to the majority of cities on the Spanish Mediterranean coast, resulting in "urban sprawl". Urban planning is characterised by the following:

- Inefficient territorial structure: populations reside in different urban centres distributed among the territory; here low-density developments stand out, along with the segregation of land use. There is no favourable infrastructure network that permits sustainable mobility, and no infrastructure unites the dispersed centres within the territory, which generates a high dependence on private cars.
- The "barrier effect": the problematic presence of significant transportation infrastructure that divides the territory and generates a considerable "barrier effect" which aggravates the present lack of infrastructure.
- Low number of trips on foot: there is a low modal quota of trips on foot in general terms. Although it is understandable that the percentage of trips on foot in this type of city cannot reach the same levels found in municipalities with a compact territorial model, these percentages reveal an opportunity for improvement.
- The limited presence of bicycles in municipal mobility: most municipalities on the Mediterranean coast present an orography and territorial structure that

suits the use of sustainable transportation such as the bicycle, even though bicycles have a limited presence in the modal split.

- A lack of urban transportation services: lack of regular urban bus service that connects the population centres detracts from the use of this mode of transport, preventing the empowerment of other sustainable modes through the promotion of intermodality.
- Underutilisation of the existing railway services: the railway infrastructure is underused, offers limited services and railway stops are difficult to access by pedestrians. In addition, the railway system doesn't connect effectively to other sustainable modes of transportation, thus causing this service to remain underused. In reality, the railway system holds great potential as an alternative to the private car to connect adjacent urban centres and nearby metropolitan areas.

Intervention strategies for improvements in mobility in the cities of the Spanish Mediterranean

This section raises proposals to improve the urban and regional mobility of the municipalities of the Mediterranean littoral and to establish "green corridors" to encourage residents' sustainable mobility. These measures should be indicative of the goals of supra-municipal agencies both for sustainable development in general and sustainable mobility in particular. In this frame, the United Nations organisation has recently published and approved the document *Transforming Our World: The 2030 Agenda for Sustainable Development* (United Nations, 2015), where 17 fixed goals are established. The context of this document was to describe how governmental organisations must act for years to come. The following are the 2030 sustainable mobility goals:

- Develop quality, reliable, sustainable and resilient infrastructure to support economic development and human wellbeing.
- Provide access to safe, affordable, accessible and sustainable transport systems for all, improving road safety, notably by expanding public transport, with special attention to the needs of those in vulnerable situations, women, children, persons with disabilities and older persons.
- Enhance inclusive and sustainable urbanisation and capacity for participatory, integrated and sustainable human settlement planning and management in all countries.
- Reduce adverse per-capita environmental impact of cities by paying special attention to air quality and municipal and other waste management.
- Provide universal access to safe, inclusive and accessible green and public spaces, in particular for women and children, older persons and persons with disabilities.

Gathering inspiration from these collective strategies, four specific proceedings using the case of L´Alfàs del Pi are offered next.

Configuration of main corridors for the structuring of the territory and the promotion of sustainable mobility

This action is defined as the refurbishment of the main roads connecting the most important population centres. This action also defines the refurbishment of the vertebrate axes that run in the city and may take a large part of the present and future mobility. The objective of these enhancements is to favour sustainable mobility in the municipality (Ayuntamiento de L'Alfàs del Pi, 2016). This goal will be attained through specific points such as narrowing road traffic lanes and constructing spacious sidewalks or bike lanes to promote cycling and pedestrian mobility in a secure environment. Moreover, it is highly recommended to include enough wooded annex to these roads to protect both pedestrians and cyclists from the sun and the heat, especially in summertime.

Applying these goals in the case of L'Alfàs del Pi is a must. First, two corridors on the main road connecting the town centre of L'Alfàs with L'Albir (both Serra Gelada and L'Albir beach) must be established. The corridors will allow for a restructuring of the municipalities' main population centres, thus favouring soft transportation modes. The corridors that would improve urban mobility are:

- Axis L'Alfàs–Playa de L'Albir: formed by the roads Camí de la Mar, a small section of Avinguda de L'Albir and Boulevard dels Músics, and connecting the town centre of L'Alfàs del Pi with L'Albir beach.
- Axis L'Alfàs–L'Albir Serra Gelada: formed by CV-763 and CV-753 or Camí vell d'Altea from crossing Avinguda de L'Albir and Boulevard dels Músics.

As shown in Figure 9.3, both corridors would remain connected by both ends. This would be carried out by implementing the green urban corridors, with lanes of 2.75 meters wide for motorised traffic. In addition, the rail must include cycling paths and wide sidewalks, allowing pedestrian and cyclist mobility in safe conditions to guarantee an alternative transportation mode to motorised vehicles. All this should be accompanied by the important presence of vegetation along the corridors, generating protection and cover from high temperatures for the most exposed users (pedestrians, cyclists, etc.) and increasing the biodiversity in the town, thus establishing authentic "green urban corridors".

In addition, these "green urban corridors" would fulfil two other functions:

- They would serve to improve the transverse permeability of the town, reducing the barrier effect exercised by the national road N-332.
- They would connect with both TRAM stations that offer service in the surrounding environment, hence favouring intermodality between cyclists, pedestrians and urban and interurban public transport users. This last idea connects with the following proposed action.

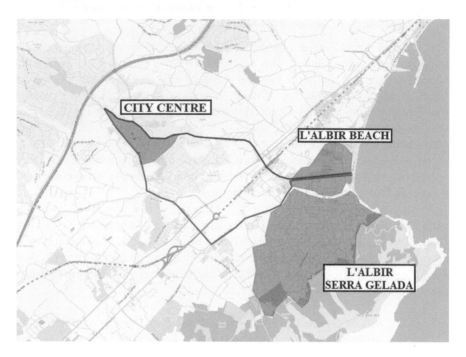

Figure 9.3 Green urban corridors proposed in the first action.
Source: Prepared by the authors with data from the OpenStreetMap (2018).

Implementation of the TOD model on the railway stations

The model called transit-oriented development (TOD) consists of raising urban development projects linked to existing stations of public transportation. In this way, the use of sustainable transportation will be intensified (City and County of Denver, 2006; Curtis, 2012). This model has been successfully implemented both in zones of low-density and compact city zones as well as in in the cities of Portland, in the United States; Karlsruhe, in Germany; and the Netherlands (Ortuño, Fernández & Fernández, 2017). Consequently, this model is well suited for medium-sized cities of the Mediterranean littoral. In the Mediterranean littoral, there exists habitually a combination of urban development models with predominance of low-density in the coastal centres.

In its application in the municipality of L'Alfàs del Pi, the TOD model would allow concentrating TRAM railway stops in urban developments in upper-middle-density ranges by utilising a mixture of land (residential, commercial, tertiary, etc.) and an urban environment very different to the current one. The new railway stops would be developed in low-density areas and through segregation of land use (Ortuño & Casares, 2017). These newly built spaces can turn into important nodes of urban connection that encourage intermodality and sustainable mobility.

In the case of the stops of L'Albir and L'Alfàs del Pi, Line 9 of TRAM, the area of influence to analyse would include a radius of 500m around each station. The focus would be on finding propitious conditions for the application of TOD, because 52% of the users that access the network are located in these areas, according to the information gathered in the last Index of Satisfaction to the Client (ISC) in 2017 (Fundación de los Ferrocarriles Españoles, 2018). The territorial space used to develop TOD, where an important part of future urban development should be concentrated, is shown in Figure 9.4. Consequently, the town hall should link actions and initiatives in the context of these urban environments.

One of the strategies implemented by municipalities is to increase tertiary land supply to attract innovative businesses, especially those working in the health field, which the municipality already specialises in. In order to complete the TOD model and facilitate attraction of innovative activities, this tertiary land must be accompanied by a commercial and residential correspondent to allow an appropriate mixture of uses and general urban activity.

Certainly, these new areas of development would also benefit from connecting the coastal zones with the interior, which would provide continuity to the urban structure. It would be ideal to construct nodes connecting L'Alfàs del Pi's urban

Figure 9.4 TRAM railway stops and areas of influence for the implementation of TOD.

Source: Prepared by the authors with data from the OpenStreetMap (2018).

area and L'Albir's urban cores, since both nodes are located between the urban centre of L'Alfàs del Pi and the urban centre of L'Albir, as seen in Figure 9.4. To be noted are the high accessibility levels that both nodes present, not only for TRAM railway stops, but for the vicinity of the national road N-332. In addition, the "green urban corridors" and bus lines that are proposed later in this chapter will cross these areas of influence. These proposals will have the greater effect of opening up possibilities to generate poles of intermodality between TRAM, bus, bicycle and walking commutes.

Pedestrianisation of urban roads and other complementary goals to encourage sustainable mobility and refurbish public spaces

One of the main consequences of low-density urban development is excessive dependence on private cars. Dependence on private cars affects residents' mobility even in urban areas. This situation causes an excessive appropriation of public and community space by the city in order to accommodate motorised traffic, which prevents pedestrians from enjoying this space in the community. It is therefore indispensable to apply a series of goals which allow citizens to occupy certain city public spaces. Pedestrians must receive priority in town centres for mobility purposes. It is necessary to plan for the pedestrianisation of various zones with more urban activity.

In the case of L'Alfàs del Pi, modal split shows a significantly low quota of trips on foot, 22% in total. Reasons for such low percentages rest fundamentally on the current disconnection between the different population centres of the municipality and the development of "urban sprawl". An element that strengthened this result is linked to the fact that the urban road and its traffic block access to pedestrians, thus putting pedestrians mostly in a secondary role. A very eloquent figure that justifies this fact is that, as discussed in the *Study of Traffic and Mobility* of the General Plan of Urban Arrangement (Ayuntamiento de L'Alfàs del Pi, 2015), there are only five pedestrian streets in this municipality, out of a total of 199 streets.

Although this measure will need to be backed up by a corresponding analysis to choose the best formula for each environment, the dimensions of L'Alfàs del Pi's town centre allow mostly for the pedestrianisation of the town centre, with the exception of the primary roads. In L'Alfàs del Pi, more than 400 "park and ride" type of parking places can be found. "Park and ride" slots are located in wide parking lots usually found on the periphery of a city and near public transportation stops. These parking places could consolidate a great part of the demand for parking, consequently eliminating parking zones on the street surface. In this way, pedestrianisation would be favoured and parking supply would not lack.

In the case of L'Albir's centres, the situation is very similar, in that up to 500 "park and ride" lots can be found. These parking spaces might allow pedestrianising important roads such as Avinguda de L'Albir and Avinguda Oscar Esplá. These two roads represent the epicentre where most of the commercial leisure activities happen in L'Albir. The municipality must consolidate the rest of the urban area and analyse the possibility to pedestrianise or to limit the access to traffic in favour of pedestrian mobility.

The improvement of all these public spaces would have a significant positive impact on the daily life of the inhabitants of L'Alfàs del Pi. Social relations between citizens would increase and, in turn, positive synergies with tourism, small business and retail could be created. In those cases where total or partial pedestrianisation isn't possible, diminishing traffic becomes necessary, as well as expanding sidewalks, establishing speed limit routes such as Zone 30 and adding bicycle rails. Complementary goals should allow democratising the use of public spaces and should eliminate the current hegemony of private cars.

Establishing regular public urban transportation

Public urban transportation (principally bus) complements transportation by bike or foot. It is necessary to reinforce the frequency of these services in order for them to become a real alternative to cars, especially in times of the day with major mobility demands. It is necessary for urban services to connect with the principal urban transportation infrastructures present in the municipality. Favouring intermodality is crucial, because most public transportation sources, including railway stations, attract a bigger user demand.

When we look at L'Alfàs del Pi, we notice that it is composed of 12 urban centres disaggregated by the territory. Because of the dispersing effect of "urban sprawl", public urban transport must play an indispensable role in order to turn L'Alfàs del Pi into a sustainable and improved territory that favours social cohesion. This last action will only be effective if it happens concurrently with the rest of proposed actions, such as building a "green corridor" and creating more pedestrian walkways. Until a homogeneous implementation happens, the use of public transportation will remain residual, as it is currently.

Urban bus services must follow a route that covers both urban corridors, thus generating synergies between different sustainable transportation modes. Bus routes should also connect to the urban centres of the zones northwest of the municipality that are located beyond the highway. In turn, this service will have to connect with two intermodal nodes generated in the environment of two TRAM railway stations, thus allowing the inhabitants of the municipality to rapidly connect with these new urban developments tied to TOD.

These actions would entail a structural change in the mobility of the municipalities in favour of sustainable modes. The proposed solutions would also generate opportunities to develop public spaces for the inhabitants of the zone to utilise. Overall, new commercial developments and manufacturers would have a chance to grow and a better utilisation of the existing infrastructure would be implemented. With the concretisations of these proposals, a better environment for residents and tourists would be achieved.

References

Ayuntamiento de L'Alfàs del Pi. (2015). *Estudio de tráfico y movilidad del Plan General de Ordenación Urbana de L'Alfàs del Pi [Traffic and mobility study for the urban plan of L'Alfàs del Pi City]*. Spain: L'Alfàs del Pi.

Ayuntamiento de L'Alfàs del Pi. (2016). *Estrategia de Desarrollo Urbano Sostenible Integrado (EDUSI) 2014–2023 [Integrated sustainable urban development strategy 2014–2023]*. Spain: L'Alfàs del Pi.

City and County of Denver. (2006). *Commuting planning & development, transit oriented development strategic plan*. Denver. Retrieved from http://ctod.org/pdfs/2006TODStrategicPlanDenver.pdf

Curtis, C. (2012). Delivering the "D" in transit-oriented development: Examining the town planning challenge. *The Journal of Transport and Land Use*, *5*(3), 83–99.

Ferrocarrils de la Generalitat Valenciana. (2018). *Moverse con TRAM – Horarios. TRAM Metropolitano de Alicante [Moving with TRAM – timetable. TRAM metropolitan railway of Alicante]*. Retrieved from www.tramalicante.es/horarios.php?page=143

Fundación de los Ferrocarriles Españoles. (2018, January 11). Más de la mitad de los usuarios del Tram de Alicante accede a la red en un radio de 500 metros [More than a half of TRAM users accesses the rail network within a radius of 500 meters]. *Vía Libre*. Madrid. Retrieved from www.vialibre-ffe.com/noticias.asp?not=23014&cs=oper

Instituto Nacional de Estadística. (2016). *Nomenclátor: Población del Padrón Continuo por Unidad Poblacional [List of place name: Population of the continuous municipal register by population unit]*. Madrid. Retrieved from www.ine.es/nomen2/index.do

OpenStreetMap Contributors. (2018). Retrieved from https://planet.openstreetmap.org

Ortuño, A., & Casares, J. (2017). Historia de los ferrocarriles regionales en la Comunidad Valenciana: una visión territorial [Regional railway history in the Valencia region: A territorial perspective]. *Transportes, Servicios y Telecomunicaciones*, *34*, 127–157.

Ortuño, A., Fernández, G., & Fernández, P. (2017). El Modelo TOD (transit-oriented development): estudio de casos internacionales y proceso de implementación [TOD model: Study of international cases and implementation process]. *Boletín de la Asociación de Geógrafos Españoles*, *73*, 99–121.

United Nations. (2015). *Transforming our world: The 2030 agenda for sustainable development*. The General Assembly 2015. Retrieved from https://undocs.org/A/70/L.1

10 Preliminary assessment practices for the definition of tools and goals of landscape quality in Mediterranean settlements

L'Alfàs del Pi as a case study

Francesco Carlo Toso

Introduction

The agricultural landscape tradition of Mediterranean littoral regions plays a fundamental role in defining a level of extensive landscape quality for their sustainable territorial development. Many regions derive their specific and historically consolidated landscape structures from the agricultural management of their natural resources. Over the centuries, water irrigation systems have been developed, morphological structures such as terraces have been built and vegetation patterns have been formed by using particular cropping techniques. In the context of the current model of territorial development, the landscape qualities are threatened by the pressure of real estate and its connected infrastructure, but their recovery and enhancement is a strategic element of territorial regeneration. The pressure exerted from urban development on littoral regions imposes urban structures, settlement typologies and land uses that fragment the landscape and results both in a loss of place specificity (Calcagno Maniglio, 2016) and in negative environmental impact (Spanish Ministry of Environment, Rural and Maritime Affairs, 2006).

Sustainable development, on the other hand, necessarily includes the dimension of cultural sustainability, as stated in the Ljubljana Declaration on the Territorial Dimension of Sustainable Development (CEMAT, 2005). The specificity of place as an element of cultural identity promotes territorial sustainability. The approach to landscape adopted here follows the principles of the Council of Europe (2000), acknowledging all landscapes as an important element for the life quality of the population, including not only areas of exceptional quality but also extending landscape assessment to areas bordering the built settlements and to the country-side. The assessment methodology considers landscape as a cultural product of the interaction of human beings with their environment and natural resources over time. It is therefore not only limited to the landscapes that are today considered most attractive and integrated into the everyday experience of the population, but extends to the marginal and residual areas. Attention is given to the characteristic long-term historic landscape qualities that pre-date the often disruptive developments brought on many Mediterranean regions by the tourism-oriented urbanisation of the latest decades. Rural landscape and its heritage are particularly vulnerable in the context of coastal tourism.

According to ICOMOS-IFLA's *Principles Concerning Rural Landscape as Heritage*, the rural landscape

> encompasses physical attributes, the productive land itself, morphology, water, infrastructure, vegetation, settlements, rural buildings and centres . . . as well as wider physical, cultural, and environmental linkages and settings . . . includes associated cultural knowledge, traditions, practices, expressions of local human communities identity and belonging, and the cultural values and meanings attributed to those landscapes by past and contemporary people and communities.
>
> (ICOMOS-IFLA, 2017)

The case study of L'Alfàs del Pi presents many of the characteristics and themes that are common to littoral tourism landscapes as well as the goals that are common to Mediterranean regions as mentioned in the Sevilla Charter of Mediterranean Landscapes (1992). A particularly important goal is to safeguard historical value while managing quality landscape transformations, involving sustainable planning of tourism infrastructure in coastal areas (Spagnoli, 2013). The methodological basis for the proposed preliminary analysis is a systemic approach to rural landscape assessment, highlighting relations between its components, taking into account the time dimension and linking of tangible and intangible values, objective information and cultural perception (Scazzosi, 2018). This methodology is informed by the best practices developed in the Pays.Med.Urban European interregional project on landscape in Mediterranean urban areas (Manzato & Poli, 2011; García & García, 2011), as well as the practice of participatory rural assessment (Chambers, 1994). The analysis involves the review of available documentation, interaction with local people and outsiders from different sectors and disciplines through participatory activities and fieldwork in the landscape.

Remaining traces of historic landscapes: the landscape of L'Alfàs del Pi

Place knowledge is the basis of any process of landscape assessment and must be based on a "knowledge project" comprising the characterisation of past and present physical and cultural features, reading the dynamics and systems that generate the landscape as well as the social and cultural perception over time (Scazzosi, 2011). When approaching the landscape of a new region for the first time, we need to collect information on the main elements that historically structured the territory over time. Despite having often lost in part or completely their functional context, those elements continue to shape our current perception of the landscape (Küster, 2012).

The climate and geological conditions of the area of L'Alfàs del Pi historically determined a rural landscape based on pluvial dryland farming (*cultivo pluvial secano*) and territorially localised irrigated orchards (*huertas*). These two cultivation systems originate two dominant elements of the local rural landscape. Dryland farming developed to enhance the water retention capacity of the soil and brought the practice of terracing on the hillside areas to gain larger surfaces for cultivation purposes. The development of techniques to maximise the use of the scarce water resources and extend the irrigation of orchards has created, over the centuries, the most important anthropic elements of the landscape in L'Alfàs del Pi. Literature suggests that at the very heart of the toponym "Alfaz" or "Alfas" is the Arab word *al-fahs*, meaning "sown field" (Asín Palacios, 1940). The beginnings of irrigation in the valleys of the river system Algar-Guadalest date back to the 11th century, during the Islamic period, when these territories saw a steady increase of the population, which stopped only with the end of the domination of the Islamic government in the early 17th century. These centuries produced the backbone of the irrigation networks that were then inherited and transformed with the creation of new channels and mills starting from the 17th century (Hermosilla Pla, 2016).

The irrigation system Reg Major de l'Alfàs y Benidorm dates back to the mid-17th century, providing water to the neighbouring lands of Polop, La Nucia, L'Alfàs del Pi and Benidorm. Promoted by the Baroness of Polop as a way to increase the productivity of her possessions and designed and directed by the architect Francisco Serrano, it comprised a main irrigation channel, aqueducts and tunnels, secondary channels and mills. The irrigation system brought significant improvements to an area that suffered from very scarce agricultural productivity and low population. Later, towards the end of the 19th century, the irrigation system was operated from a community of *regantes* (irrigators) and received proper usage regulations in 1927, although by this point its importance was largely surpassed by newer irrigation works using the waters of the Algar river. Nevertheless, the Reg Major infrastructure has left many artefacts that still exist and deserve inclusion in the landscape planning, some clearly visible, like the aqueducts (Acueducto de Soler, de Carbonera and Arcs), and therefore classified as highly valuable heritage, some less noticeable, particularly the tunnels, the numerous linear channels with their different designations and purposes as well as numerous irrigation ducts (Fernández, 2015). Of particular importance, not least for being still mentioned in the local toponyms, are the *molinos*, the mills activated along the Reg Major, mainly for flour production (Fernández & Julián, 2015). Literature suggests that the heritage water utilisation network, whether active or discontinued, holds the most representative identity value at the local level, and is the "most complete expression of the water cultural landscape" (Hermosilla Pla, 2015).

The contemporary landscape and the limits of urban development

It is a common phenomenon for Mediterranean landscapes shaped by stable long-term processes to have suffered from quick territorial changes over the course of the second half of the 20th century. It is therefore essential to frame these changes and their impact on the historic landscape structures. The territory of L'Alfàs del Pi has been examined by researchers from a mainly sectoral geographical and historical perspective as part of the larger study area of the Marina Baixa, predominantly under territorial and functional relations. Prevailing themes are the territorial consequences of the economic changes in agriculture as well as of seasonal and residential tourism on water resources (Hernández Hernández, 2013). Agriculture in the region changed from the formerly prevailing dryland farming cultivations to intensive irrigated and the ploughing of the shrubland areas. Agricultural production is predominantly export-oriented and is particularly water-resource intensive; smaller, local-scale agriculture tends to be abandoned due to the prevailing use of local water resources for residential settlements, as well as its low profitability (Ortiz, Inmaculada, & Melgarejo, 2016). An absence of adequate planning and governance tools, political interest and public awareness often results in a total change of land use, which produces the gradual loss of the *huerta* landscape (Romero & Melo, 2015).

Conflicts also arise by hosting agriculture and tourism in the same areas. The tourism sector shows two distinct development models. First, seasonal seaside tourism, which determined the urbanisation boom of the coastline, and whose peak is represented by the high-density and high-rise settlement of which the city of Benidorm, neighbouring L'Alfàs del Pi, is the prime example. Next to it we observe a semi-permanent model, where foreign tourists look for a more diverse quality of land to purchase in their surroundings, defined by lower density, more greenery or a more picturesque, varied and less urban character including not only the coastline but different elements, such as the mountain enclosures and the hilly countryside scenery that are found in L'Alfàs del Pi. It's this second type of tourism that, supporting both the construction sector and municipalities through revenues and the flourishing of dedicated activities, determines a more sprawling kind of urban development, paradoxically undermining exactly those characteristics that are sought in the long-term. On the other hand, the switch in agricultural demand determines the use of greenhouses and protection nets which have a significant visual impact in the mid-range landscape due to their large uniformly textured and reflective plastic surfaces covering large land plots (Campagne, Cantó López, Hernández Hernández, & Llopis, 2005).

Useful information on the trends that modify the agricultural landscape can be retrieved by analysing the development of parcel structures. In our case study area, the growth of urban settlements starting from the 1970s brought a reduction of the crops in the Marina Baixa region of about one-third in the years 1956–2008, a trend which was especially strong in the early 2000s (González Ferrarió & Hermosilla Pla, 2015). Interestingly enough, this trend, which appears to be significant in many municipalities, does not appear to have significantly affected L'Alfàs del Pi up to the last decade. In fact, in 2008 the totality of traditional irrigation was operational and overall 70% of the agricultural parcels within the municipality borders were small, compared to the larger cultivated parcels in the Marina Baixa.

The latter development mode was deemed ecologically unsustainable due to the parallel factors of annual water deficit, the connection between high tourism and uncontrolled urban development, high demographic concentration in the coastal and intermediate zones and the increasing water use in intensive irrigation crops. The increased water use generates negative water balances, particularly in the intermediate and coastal region of the Marina Baixa, where L'Alfàs del Pi is also situated (Chirino Miranda, Abad Chaves, Abad, & Francisco, 2008).

If the Reg Major cuts the territory of L'Alfàs del Pi from north to south in the direction of Benidorm, flowing across the inland, an equally defining element of the water landscape are the two gorges (*barrancos*) flowing towards the coastline, constituting two natural "green corridors". The area of the Barranco de Soler is connected with the old town of L'Alfàs del Pi and the L'Albir coastal settlement, while the Barranco de Carbonera, of a more rural landscape quality, flows through an area of low urban density and offers a potential connection with the territories of Polop and Altea. Local planning already acknowledged the potential of the two *barrancos* to constitute a true green infrastructure connecting the uphill landscape with the coast, through appropriate maintenance and safe accessibility of what today are considered marginal and residual areas. These plans are meant to set off general planning objectives, namely the consolidation of a green infrastructure connecting the interior and the coast while preserving the environmental and cultural heritage of the area. The territorial strategy furthermore suggests the creation of a cluster of areas where new economic territorial uses are to be experimented (Generalitat Valenciana, Estrategia Territorial, 2010–2030, Área Funcional de La Marina Baixa).

Assessing the landscape: research hypothesis and methodology

As a starting point for the landscape assessment we also need to understand whether the resident population is aware of the overall quality and potential of the local landscape. In fact, landscape is not only the collection of a number of natural and territorial features but includes and is actively shaped by how its inhabitants, users and agents perceive it, as suggested by the European Landscape Convention. The foreign population may be mainly interested in the features of a "recreational landscape", connected with the prevailing tourism-oriented offer. The local population, although aware of the main traditional features, has experienced the loss of landscape structure and the predominantly tourism-based use of the region. Locals have mostly no contact with the town's primary sector and remain largely unaware of the landscape's ample value, which has been acknowledged by researchers as integral to the identity of the settlement (Frías Castillejo, 2014).

Whether the actual perception reflects this working hypothesis could obviously not be verified via bibliographical research, available cartography and mapping. Furthermore, an *ad hoc* assessment over time is needed. In particular, it is practically impossible to assess perceptive qualities of the landscape at a distance and characterise areas that are mostly not portrayed by any available media or bibliographic source. The available cartography allows for a comparison of the past and contemporary land use that can only serve as a starting point for an *in situ* exploration.

The initial assessment was carried out in three steps, leading to a first definition of existing "landscape values" as well as elements of potential development:

*   Analysis of the early available planning, cartographic and photographic documentation
*   Interviews with the citizens and stakeholders taking part to the participatory workshop
*   On-site assessment, with an early exploration of the territory

The mapped landscape: L'Alfàs del Pi in cartography and planning documentation

An overview of the available cartographic documentation provides a direct comparison of the most recent change in landscape morphology and land use between the 1950s and today. Although based on 2D aerial views, this comparison allows us to assess which areas consistently retained their landscape characteristics and which areas were subjected to an almost complete transformation through urbanisation and infrastructure.

The earliest aerial photos available were taken from the U.S. Army Map Service in 1956–1957 in the context of the agreements with the Spanish regime during the Cold War (De la Fuente, 2016) and provide us with a clear picture of what the area of L'Alfàs del Pi must have looked like before the touristic boom of the 1970s and 1980s. A striking feature is extensive terracing of the hillsides to host dryland cultivations. Also clearly identifiable are the more densely planted areas, the *huertas*, as well as the old town of L'Alfàs del Pi with its historical roads and the clear infrastructural sign of the railway cutting the uphill settlement from the coast, which influenced significantly the internal division of the territory up to the present day.

A more in-depth study of the landscape's historical characteristics requires a systematic analysis of its macro-structures, with a thorough mapping of the road systems, plot subdivision and terracing systems at subsequent time steps. Using the available aerial photography as a basis to conduct a proper archeo-morphological study of the historic landscape would be beneficial. Information retrieved from aerial photographic sources can be also connected with the study of cadastral documentation – there are examples of this technique applied to the study of terracing systems in the Alicante province and elsewhere in Spain (Riera Mora & Palet Martinez, 2008; Asins Velis, 2011). This would allow us to render the stratification in the anthropic morphological structure and assess the historical value and the integrity level of the surviving landscape structures. However, for the purpose of an early assessment, a simple comparison of the aerial photography is sufficient and allows us to subdivide the study area in three groups: land use not changed from agricultural to urban, with clearly recognisable morphology; land use changed, while the morphology is still clearly recognisable; and permanence of historic settlements. This simple classification allows a first evaluation of the integrity of historic landscape as a working hypothesis.

An overview of the local planning documentation brings awareness around the results of previous analyses as well as their limitations. The general plan of L'Alfàs del Pi includes a Catalogue of Landscapes. The catalogue was compiled with the intention of "defining the goals for a sustainable development where development and preservation of landscape values coexist" and "identifying the environmental, cultural and visual traits that are valued by the population" (Prieto Cerdán, 2015a, p. 2). The catalogue includes only two "Landscape Units" subjected to the law for the protection of natural spaces, namely the whole Serra Gelada natural reservoir and the orchard area named El Planet, while single localised items are presented as "Landscape Resources" (L'Albir lighthouse, the Torre Bombarda fortifications, the church square in L'Alfàs del Pi, L'Albir beach, the two gorges and the archaeological remains of Villa Romana). These landscape units and localised resources are assessed with additional criteria of "visibility" – taking into account purely visual perceptive criteria – and "landscape value", which again identifies a "visual quality" (Prieto Cerdán, 2015a, p. 11). Interestingly, the visual analysis considers landscape perception from short, medium and long distances and takes into account only "dynamic" observation points, stating that for the topographic characteristics of the territory, "no panoramic point can be considered" (Prieto Cerdán, 2015b, p. 51), although any of the settlements and paths in a panoramic position may be used to analyse "static" or slow mobility perception. Instead, the local planning landscape study chooses seven paths, mainly corresponding with roads where the landscape is most often perceived from the point of view of travelling motor vehicles.

The landscape quality goals in local planning are thus built on an assessment of the natural and ecological resources and on a restricted listing of items of cultural importance. The rural areas and activities and the historic structure of the landscape are not explicitly mentioned as a goal of landscape quality, other than in a paragraph mentioning "new economic activities based on qualifying the territory and innovation". El Planet represents here the main rural landscape unit, where cultivated plots prevail; it borders the municipality of Altea, but it's not the only one. In particular in the Landscape Unit named UP3, around the main urban settlement of L'Alfàs del Pi, bordering Benidorm, areas with citrus cultivations are present, along with many abandoned fields (see Figure 10.1).

Consistent with the approach of landscape characterisation (Pérez, 2009), the study focuses on the natural and anthropic elements that structure the landscape. A scoring system is used to categorise landscape quality, based on "scene quality" (*calidad de la escena*). It is based on a "physiographic" criteria, where a higher quality is assigned to higher structural form complexity. "Vegetation" criteria are assigned, evaluating the diversity and quality of cultivations, tree bushes and shrubland. Other elements taken into consideration are the presence of anthropic artificial elements, the composition and the "relevance of the scenery", which highlights the most representative places. The authors use these criteria to assess

Figure 10.1 The agricultural area UP3 in L'Alfàs del Pi, a peri-urban agricultural space
bordering the high-density neighbouring city of Benidorm.

seven identified landscape units, with the result that the areas of the Serra Gelada
and El Planet are classified as having a high landscape value, the Cautivador a
middle value and the urban settlements of L'Albir, the old town of L'Alfàs del Pi
and Escandinavia a low landscape value. The integration of the rural character of
the municipality is mentioned in the landscape assessment for the municipal plan
but is not given any further detailed analysis. The integration of rural characters
into green infrastructure remains at a very generic level since no further formu-
lation is found and it is unclear which tools could be deployed for the eventual
preservation of rural landscape (Prieto Cerdán, 2015b, p. 92).

Two topics appear to need further exploration in order to gain a more complete
picture. Perception-based analysis needs to be completed with observations on
slow mobility perception that go hand in hand with the need to develop a slow
mobility network for tourism. These observations can also be deployed as useful participation activities. The perceptive approach cannot be mainly visual but
should be multi-sensory and clearly include the cultural aspects. Cultural perception is in fact grounds for active participation of the population in landscape
making. One should not forget that landscape includes subjective and cultural
components which can only turn into a resource if the local population takes on a
role as active "landscape makers", which means being aware of the landscape and
becoming involved in using and maintaining it. A clear reference to the historic
rural landscape structure and its value is needed to fully realise the potential of

an integrated "landscape infrastructure", not only to connect "green" areas, but as a basis to create a network of "landscape actors" and policies to support the infrastructure over time.

The insiders' perspective: participation activities and cultural perception

To test these initial considerations with the point of view of the "landscape insiders", a preliminary "cultural perception" activity was organised as part of the larger participation process. During a workshop held in June 2017 in the context of a study commissioned by the municipality of L'Alfàs del Pi, participants were asked to take part in a quick session on the topic of their subjective perception of the landscape. The participants were divided in four work groups and asked to list places or particular symbols, buildings or images on three sets of coloured sticky notes. Each of the three colours was associated with a particular theme:

1   Places, views or symbols associated with identity, tradition and memory
2   Places, views or symbols that are most important on a personal level
3   Places that are perceived as contradictory, problematic or extraneous

The participants were given a very short time and were encouraged to write spontaneous associations. The goal was to gain early insight into significant elements of the landscape that may not be otherwise highlighted in the existing literature and that may not be evident to an "outsider". The focus was not on collecting geographical, environmental or architectural information, but on collecting a first impression of the cultural or even emotional experience of the place by its inhabitants. Emotions often bring attention to salient environmental features. By collecting emotional associations, we may get a sense of which places retain collective memories (Duncan & Duncan, 2010).

The participants included local inhabitants, mostly native but also long-term residents from abroad, representatives from local associations, local administrators and administrative technicians. The majority of the participants can thus be considered "landscape insiders", according to a definition introduced by geographer Denis Cosgrove (1984) to indicate the population whose perspective on a particular landscape is shaped by living in direct contact with it, as opposed to seasonal tourists as well as to the research team, who are "landscape outsiders". The insider view is important because it is shaped by long-term interaction with the landscape and informed by belonging to the local community, whereas the outsiders' view is, at least at first, mostly based on perception and its interpretation through an external cultural background (Turri, 1998).

Here is the list of answers that were given by workshop participants:

1   Places, views or symbols associated with identity, tradition and memory:

the church square and the pine in the church square of L'Alfàs del Pi; the Albir lighthouse; L'Albir beach; *las acequias*, the old irrigation channels; the old city (*casco antiguo*) of L'Alfàs del Pi; the weather, the natural setting between sea and mountains; a sketch of an aqueduct bridge; fireworks and traditional processions; the square; agriculture; Cala del Amerador cove; the traditional folk festivity with chants and dances held in the old town of L'Alfàs.

2   Places, views or symbols that are most important on a personal level:

L'Albir beach; "el Tossalet" district; L'Albir lighthouse; the Casa de Cultura cultural centre; the Fundacion Frax private cultural centre; "market, cafes, restaurants"; the fields, the sea, the natural features; the cove at the old mine of L'Albir, a small sketch made by a participant depicts sea, mountains and sun.

3   Places that are perceived as contradictory, problematic or extraneous:

the national road; the hotel and clubs in the Ventorillo district; ruined houses in the *casco antiguo*; abandoned terrains; trash abandoned in the fields.

We can immediately identify the *iconemes* of the landscape. The concept of *iconeme* is borrowed from the discipline of semiology and applied to the landscape. According to Italian geographer Eugenio Turri (1998), an *iconeme* is an "elementary unity of perception" (p. 170) that isolates a representative portion of the landscape and builds the syntax which we recognise as landscape. A major *iconeme* can be associated with the idea of a landmark, an element whose range of inter-visibility and whose predominant characteristics help to identify and sum up a landscape (Turri, 2010). Two *iconemes* in particular are most representative of the local identity: L'Albir lighthouse in the Serra Gelada natural reservoir (see Figure 10.2) and the small church square with the pine tree in the old city of L'Alfàs del Pi. Interestingly enough, the small folk museum in the old city of L'Alfàs del Pi receives no mention.

Identity is associated with history and tradition; in particular, the festivities which take place mostly in the old city of L'Alfàs del Pi seem to play an important role. There is also a clear reference to the former agricultural character of the area and the historic irrigation system. The area's more tourism-oriented, seaside resort character is hardly mentioned as part of the "core identity" of L'Alfàs del Pi, thus confirming the strong division between the historic town and the tourism-oriented settlements. A different picture emerges from the list of places that bear a stronger personal association with individual memories: the places of recreation and free time are strongly represented, as well as places that are shared with tourism. Places which have not been mentioned as symbols of identity are mentioned as an important part of personal everyday life.

The national road is clearly and immediately identified as problematic, coherently to what emerges from the other workshop activities. Equally problematic are the abandoned terrains and degraded fields, as well as the ruined houses in the old town. Care for the historic urban and rural landscape is a prominent feature, whereas the isolated tourist settlements are perceived as extraneous.

Figure 10.2 L'Albir lighthouse, situated in a natural reservoir, has a strong symbolic and identity value which makes it an *iconeme*.

When reviewing this list, the area's agricultural past and present do not appear to feature prominently in the locals' perception; in fact, agriculture was mentioned only once in the "personal" category and not as an identity trait. Few participants seemed to place agricultural identity at the top of the list, although it is to be noted that an open reference was made to the *acequias* and the old irrigation system.

Early on-site analysis: walking through the landscape

To complement the activity with the workshop participants and as a basis for planning future possible participation activities, a brief *in situ* exploration of the

landscape is the most effective way to obtain a reality check on collected information. The walking observation should already be informed by a preliminary
analysis, to avoid relying too much on impressions and the potential "positive
disposition" that the walking practice itself produces (Macpherson, 2016). The
walking practice is otherwise an essential tool that serves as empirical evidence
of the collected hypothesis on perception, a research method with an experi-
mental character which fosters creative engagement and combines planned and
unplanned elements, revealing information which may be relevant both for design
and research (Schultz & van Etteger, 2017). Due to the involvement that the walking practice produces with the object of study and the direct interaction with the
landscape, it is one of the most effective participatory activity tools on the theme
of landscape integration, provided that social inclusion of the mobility-impaired
is taken into account.

In an early walk held in June 2017, a number of values were clearly confirmed
and a few spontaneous uses spotted. There are a few plots of land in the municipality of L'Alfàs del Pi, particularly between the more densely constructed areas,
that lack use and are not qualified as either natural or rural land. Some of these
plots of land, which belong to private owners or are advertised for sale, include
brush trees, while others are more barren and include residual dry-stonewall terraces and sparse vegetation. There are a number of country roads that are still
paved in loose gravel and retain their rural quality – in these areas, between what
looks like abandoned plots waiting for future urbanisation and residual vegetation, one can observe small-scale agriculture-related activities taking place, such
as horse or sheep breeding, as well as spontaneous uses of the country roads for
jogging and mountain-bike riding. These apparently marginal parts of the territory
do not appear to be considered part of everyday life by the locals or the tourists,
but actually bear the potential to be an integral part of the "green infrastructure"
of L'Alfàs del Pi. These on-site observations let additional landscape qualities
emerge, suggesting that further landscape-oriented participation activities are
needed to gather more close-range, small-scale insight, as well as to understand
how locals and outsiders relate to the areas that are less included in their daily use,
experience and perception.

Conclusions and suggestion for further activities

The proposed methodology shows how information from interdisciplinary
approaches can be brought together in order to outline a multi-layered reading of
the contemporary landscape and its actors. The interdisciplinary approach may
be combined with participatory elements as a basis for defining landscape quality
goals, helping point to common themes that are found in Mediterranean peri-
urban landscapes.

The observations gathered on the L'Alfàs del Pi case study highlight existing "landscape values" as well as elements of potential development. Among the
identified landscape values are morphological peculiarity, mixed-use and clear
readability of different landscape layers. Potentialities include the multifunctional

use of existing hydrographic network, the revitalisation of agricultural activities in support of health-oriented services and the interconnection of existing rural paths with slow mobility networks at a supra-local level.

The study suggests three main directions of improvement for a greater integration of the different landscape qualities in L'Alfàs del Pi. First, making the natural green corridors accessible and identifying *ad hoc* paths for slow mobility in the rural and less-qualified areas. Second, providing clear and safer slow mobility connection with routes and landscape highlights at a supra-local level. Third, increasing and supporting the landscape quality of the remaining agricultural areas through more visibility and adequate support policies and forms of governance to reduce real estate pressure and encourage agricultural continuity. It is fundamental to involve the agricultural sector and promote the multifunctional role of small-scale, local farming in maintenance of the landscape quality of rural land.

The goals need to be communicated to increase awareness on these fundamental but generally overseen characteristics of the local landscape. It is necessary to work on perception through planned participatory activities, seeking collaboration with the existing cultural structures (e.g., schools and cultural centres). A community, both local and foreign, understanding and sharing the value connected with the history and peculiarity of the rural landscape is the prerequisite to implement further measures for landscape protection and its integration in sustainable territorial planning.

References

Asín Palacios, M. (1940). *Contribución a la toponimia árabe de España*. Madrid: Consejo Superior de Investigaciones Científicas.

Asins Velis, S. (2011). *El paisaje agrario aterrazado: Diálogo entre el hombre y el medio en Petrer (Alicante)*. València: Universitat de València.

Calcagno Maniglio, A. (Ed.). (2016). *Paesaggio costiero, sviluppo turistico sostenibile*. Roma: Gangemi Editore.

Campagne, D. M., Cantó López, M. T., Hernández Hernández, M., & Llopis, J. P. (2005). Turismo y agricultura intensiva en la Marina Baixa tensiones jurídico-ambientales en zonas de alto impacto turístico. *Revista de Derecho Urbanístico y Medio Ambiente, 220*(39), 167–203. Retrieved from https://libros-revistas-derecho.vlex.es/vid/intensiva-marina-baixa-tensiones-turistico-328494

CEMAT. (2005). *Thirteenth European Conference of Ministers Responsible for Regional/ Spatial Planning (CEMAT)*. Council of Europe.

Chambers, R. (1994). Participatory Rural Appraisal (PRA): Analysis of experience. *World Development, 22*(9), 1253–1268. https://doi.org/10.1016/0305-750X(94)90003-5

Chirino Miranda, E., Abad Chaves, J., Abad, B., & Francisco, J. (2008, April). Uso de indicadores de Presión-Estado-Respuesta en el diagnóstico de la comarca de la Marina Baixa, SE, España. *Ecosistemas, 17*(1), 107–114.

Cosgrove, D. E. (1984). *Social formation and symbolic landscape*. London: Croom Helm.

Council of Europe. (2000). *European Landscape Convention, Florence 20/10/2000, European Treaty Series No. 176*. Retrived from https://www.coe.int/en/web/conventions/full-list/-/conventions/treaty/176

De la Fuente, A. F. (2016). Los vuelos americanos de las series A (1945–46) y B (1956–57). *Andalucía En La Historia, 52*, 86–91.

Duncan, N., & Duncan, J. (2010). Doing landscape interpretation. In *The SAGE handbook of qualitative geography* (pp. 225–247). London: Sage Publications Ltd. Retrieved from https://doi.org/10.4135/9780857021090

Fernández, M. A. (2015). Las comunidades de regantes en la cuenca del Río Amadorio y sus tributarios. Gestión y uso del agua en los riegos del Amadorio, Finestrat, L´Alfàs del Pi y Benidorm. In M. Á. González Ferrarió (Ed.), *Los Regadios tradicionales de la Marina Baixa* (pp. 65–74). Valencia: Departament de Geografia, Universitat de València.

Fernández, M. A., & Julián, J. S. (2015). Los artefactos hidráulicos de la Marina Baixa. In M. Á. González Ferrarió (Ed.), *Los Regadios tradicionales de la Marina Baixa* (pp. 145–176). Valencia: Departament de Geografia, Universitat de València.

Frías Castillejo, C. (2014). Proyecto de recuperación del Reg Major de l'Alfàs y Benidorm (l'Alfàs del Pi, Alicante). In C. Sanchis-Ibor, G. Palau-Salvador, I. Mangue-Alférez, & L. P. Martínez-Sanmartín (Eds.), *Irrigation, society and landscape. Tribute to Tom F. Glick* (pp. 1105–1118). Valencia: Editorial Universitat Politècnica de València. Retrieved from https://doi.org/10.4995/ISL2014.2014.214

García, S. A., & García, A. A. C. (Eds.). (2011). *Pays.med.urban project: Catalogue of good practices for the landscape*. Murcia: Región de Murcia.

Generalitat Valenciana, Estrategia Territorial 2010–2030, Área Funcional de La Marina Baixa. Retrieved March 11, 2018, from www.habitatge.gva.es/documents/20551069/163769014/Texto+La+Marina+Baixa/ff88867f-02fc-41c0-ade6-f2beb664d478

González Ferrarió, M. Á., & Hermosilla Pla, J. (2015). Contextualiyación de los regadíos del sector meridional de la Marin Baixa. In M. Á. González Ferrarió (Ed.), *Los Regadios tradicionales de la Marina Baixa* (pp. 55–64). Valencia: Departament de Geografia, Universitat de València.

Hermosilla Pla, J. (2015). Introducción. El regadío histórico de la cuenca del río Amadorio y su entorno. Consideraciones generales. In M. Á. González Ferrarió (Ed.), *Los Regadios tradicionales de la Marina Baixa* (pp. 13–20). Valencia: Departament de Geografia, Universitat de València.

Hermosilla Pla, J. (2016). Los sistemas de regadíos tradicionales del río Algar-Guadalest (la Marina Baixa, Alicante): Patrimonio Cultural Hidráulico Mediterráneo. In J. F. Vera, J. Olcina, & M. Hernández (Eds.), *Paisaje, cultura territorial y vivencia de la geografía. Libro homenaje al Profesor Alfredo Morales Gil* (pp. 167–212). San Vicente del Raspeig: Publicaciones de la Universidad de Alicante.

Hernández Hernández, M. (2013). Análisis de los procesos de transformación territorial en la provincia de Alicante (1985–2011) y su incidencia en el recurso hídrico a través del estudio bibliográfico. *Documents d'Anàlisi Geogràfica 2013, 59*(1), 105–136. Retrieved from http://rua.ua.es/dspace/handle/10045/33111

ICOMOS-IFLA. (2017). *ICOMOS-IFLA principles concerning rural landscapes as heritage*. Retrieved from www.icomos.org/images/DOCUMENTS/General_Assemblies/19th_Delhi_2017/Working_Documents-First_Batch-August_2017/GA2017_6-3-1_RuralLandscapesPrinciples_EN_final20170730.pdf

Küster, H. (2012). *Die Entdeckung der Landschaft: Einführung in eine neue Wissenschaft*. München: Beck.

Macpherson, H. (2016). Walking methods in landscape research: Moving bodies, spaces of disclosure and rapport. *Landscape Research, 41*(4), 425–432. https://doi.org/10.1080/01426397.2016.1156065

Manzato, V., & Poli, F. (Eds.). (2011). *Paisajes de oportunidad/Evolving Landscapes. European landscape convention and participation: The pilot actions of pays.med.urban project.* Santarcangelo di Romagna: Maggioli.

Ortiz, L., Inmaculada, M., & Melgarejo, J. (2016). Evolución histórica de la agricultura de la provincia de Alicante, 1900–2000. In J. Olcina Cantos & A. M. Rico Amorós (Eds.), *Libro jubilar en homenaje al profesor Antonio Gil Olcina* (p. 1063). Alicante: Universidad de Alicante, Instituto Interuniversitario de Geografía. https://doi.org/10.14198/LibroHomenajeAntonioGilOlcina2016-56

Pérez, L. C. (2009). El carácter del paisaje y sus lectura. In L. C. Pérez (Ed.), *El paisaje : de la percepción a la gestión.* Madrid: Liteam.

Prieto Cerdán, A. (Ed.). (2015a). *Catálogo de paisajes del Plan General de l'Alfàs del Pí (Alicante).* L'Alfás del Pi: Ayuntamento de L'Alfàs del Pi.

Prieto Cerdán, A. (Ed.). (2015b). *Estudio de Paysaje del Plan General de l'Alfàs del Pí (Alicante).* L'Alfás del Pi: Ayuntamento de L'Alfàs del Pi.

Riera Mora, S., & Palet Martinez, J. M. (2008). Una aproximación multidisciplinar a la historia del paisaje mediterráneo: la evolucion de los sistemas de terayas con muros de piedra seca en la sierra de Marina (Badalona, Llano de Barcelona). In R. Garrabou & J. M. Naredo (Eds.), *El paisaje en perspectiva histórica. Formación y transformación del paisaje en el mundo mediterráneo* (pp. 47–90). Zaragoza: Prensas Universiatarias de Zaragoza.

Romero, J., & Melo, C. (2015). Spanish Mediterranean huertas: Theory and reality in the planning and management of peri-urban agriculture and cultural landscapes. *WIT Transactions on Ecology and The Environment, 193,* 585–595. Retrieved from https://doi.org/10.2495/SDP150501

Scazzosi, L. (2011). Limits to transformation in places identity: Theoretical and methodological questions. In J. Agnew, P. Claval, & Z. Roca (Eds.), *Landscape, identity, development.* Farnham: Ashgate Publishing.

Scazzosi, L. (2018). Landscapes as systems of tangible and intangible relationships: Small theoretical and methodological introduction to read and evaluate rural landscape as heritage. In L. Scazzosi & E. Rosina (Eds.), *The conservation and enhancement of built and landscape heritage: A new life for the ghost village of Mondonico on Lake Como* (pp. 19–40). Milano: PoliScript.

Schultz, H., & van Etteger, R. (2017). Walking. In A. van den Brink, D. Bruns, H. Tobi, & S. Bell (Eds.), *Research un landscape architecture methods and methodology* (pp. 179–193). Abingdon: Routledge.

Spagnoli, L. (2013). Il Mediterraneo europeo: gestire e valorizzare le trasformazioni paesaggistiche. *Documenti geografici, 0*(1). https://doi.org/10.19246/dg.v0i1.29, 10.4458/0740–07

Spanish Ministry of Environment, Rural and Maritime Affairs. (2006). *Integrated coastal zone management in Spain.* Retrieved November 20, 2016, from http://ec.europa.eu/ourcoast/download.cfm?fileID=1323

Turri, E. (1998). *Il paesaggio come teatro: dal territorio vissuto al territorio rappresentato.* Padova: Marsilio.

Turri, E. (2010). *Il paesaggio e il silenzio.* Padova: Marsilio Editori.

11 Domestic environment monitoring and its influence on quality of life

Laura García, Jaime Lloret, Ángel T. Lloret, Sandra Sendra

Habitability and standard of the dwelling

Creating a quality habitat in dwellings has been one of the most sought-after objectives since the first settlements. Nomad habitats involved somewhat different objectives than first population centres. In first population centres, the use of building systems to improve interior quality has been observed. A tendency to construct safe and protective resting spaces has also been noticed. For example, roof waterproofing and strategic orientation of façade openings were employed in first population centres. A continuous emphasis in home design and looking for commodities for resting and performing household chores has been detected as well. These home-improvement measures occur in most cultures and depend on building typologies. Our modern measures of home living standards are different in that they are regulated in order to guarantee wellbeing and comfort in dwellings.

In 1944 the Order of February 29 was published. The legislation was aimed at regulating the minimum hygienic conditions houses must have in the Spanish territory. It then became obligatory for architects to abide by the Order when drawing their designs. Medical professionals had to certify the state of the existing homes in the Spanish territory as well. The Order of February 29 determined the minimum number of rooms that should comprise a house, its minimum dimensions and other health-related aspects such as lighting type, waterproofing and insulation of the building envelope as well as the type of wastewater and infrastructure.

In 1969, the Orden Ministerial "Ordenanzas Provisionales de Viviendas de Protección Oficial" (Provisional Ordinances of Social Housing) of May 20 was approved. All houses had to abide by this law, however, the planned technical regulations for design and quality were not approved and the state functions were not transferred to the autonomous communities. In the case of the Comunidad Valenciana, the rules for habitability and design of housing (Normas de Habitabilidad y Diseño de Viviendas) were only approved later, in 1989. The Normas de Habitabilidad y Diseño de Viviendas state that it is obligatory to abide by the extended rules established in 1944 and the ones added in 1989.

In 1991, the Order of April 22 of the Conseller de Obras Públicas, Urbanismo y Transportes was approved. This document contains the consolidated text on the habitability and home design rules pertaining to the Comunidad Valenciana.

These rules are intended to define the properties that houses must have to in order to satisfy the demands of the user. Such rules are comprised of a variety of technical specifications obtained from experience and they are explained in a descriptive manner. The Order of April 22 revoked the requirements demanded from existing houses prior to the approval of the law and established new criteria for construction of brand-new ones. All conditions established by the Order applied to buildings as well as common areas.

The order of April 22 uniquely requires installing a ventilation system for enclosures or rooms. The minimum exit flow rate in mechanical ventilations for kitchens, bathrooms and toilets without outside windows is included as well. Furthermore, the type of cladding is also regulated. The regulations emphasise that certain rooms need to be hygienic and washable. The regulations also establish the hygrothermal characteristics of the materials used in construction. Sound attenuation in newly built dwellings is also discussed.

Finally, in 2009 the Order of December 7 of the Consellería de Medio Ambiente, Agua, Urbanismo y Vivienda was approved, where the design and quality conditions as mandated in Decree 151/2009 of October 2 of the Consellería are stated. The principal objective of the Order of December 7 is to develop basic design and quality requirements in residential buildings as stated in the Royal Decree 314/2006 of March 17, where the CTE (Código Técnico de la Edificación) was approved, and Law 38/1999 of November 5 where the LOE (Ordenación de la Edificación) was sanctioned.

In the CTE, the annual accumulated CO_2 concentration in residential spaces is stated. The annual accumulated CO_2 concentration now must be amended with the amount of external air flow. However, CO_2 concentrations in residential spaces are not considered satisfactory in relation to a minimal flow table or in relation to other factors such as the number of rooms of the dwelling or the type of space. The factors mentioned constitute empirical data and errors that can be produced due to changes. These errors can occur due to broken windows, badly performed ventilation, an interior redesign or not opening the windows in winter. When these errors happen, the accumulated CO_2 concentrations can vary, resulting, for example, in an increased concentration due to the lack of ventilation.

Considering the aforementioned regulation, it is of great importance to monitor indoor environments in order to ensure wellbeing. Moreover, QoL should be monitored because many factors may affect it. Currently, there is no regulation on QoL monitoring employing technology, but the first steps in this direction must be taken in order to promote future common procedures for QoL monitoring through sensors and smartphones. The rest of this chapter is structured as follows: Section 2 presents related work, Section 3 presents the proposal, Section 4 presents the results and finally Section 5 presents the conclusion and future work.

Related work

There are many proposals that employ sensors to monitor different aspects of people's everyday lives in domestic environments. Many proposals focus on older people and their needs. Bedi, Kunar and Singh (2016) performed a survey on the different sensors available for monitoring activities in smart homes. The authors identified 14 types of sensors for such purposes as monitoring power consumption, environment and air quality. The categories were motion sensors, perimeter sensors, temperature and humidity sensors, smoke and air sensors, leak and water sensors, light sensors, power synching sensors, AC/DC voltage sensors, current transformers, smart plugs, dry contact sensors, power monitoring and smart home monitoring kits.

Cavanaugh, Coleman, Gaines, Laing, and Morey (2007) presented a step-monitoring study on elderly people. The aim of their proposal was to characterise ambulatory activity in elderly people living in community dwellings. The researchers employed StepWatch 3 to gather data for six days. Results showed that healthy elderly people were less active, and their activity was less variable than younger adults. Thus, the researchers concluded that it is possible to determine the differences in activity in elderly people of different conditions employing sensors. Another monitoring study for elderly people in community dwellings is presented by Alice Coni et al. (2016), who monitored activity performed by older people employing smartphones. The system was able to determine different activity profiles. Results showed that smartphone-based activity monitoring can be employed for clinical assessment, as there has been consistency between obtained data and clinical results.

Other solutions are focused on monitoring certain environmental conditions. Chen, Chang and Chen (2015) introduced a multisensor intelligent monitoring system for home environments. They employed temperature, humidity, gas and smoke sensors to monitor the environment of a dwelling. Moreover, they employed the Wi-Fi access point in order to transmit the information. An electric fan and a dehumidifier were turned on or turned off depending on the obtained readings. Martin Leidinger et al. (2016) presented an air-quality monitoring system for indoor environments. Their system employed a gas pre-concentrator and a gas sensor. Simulations were performed in order to assess the system's performance. Finally, authors presented the gas sensor system that included MOS sensors and LEDs that showed the status display. Lita, Visan, Cioc, Mazare, and Teodorescu (2016) introduced a monitoring system of indoor environmental parameters for building automation systems. The authors employed the Atmega 328 microcontroller, an AM2302/DHT22 temperature-humidity sensor, a BMP 180 pressure sensor and an LDR for light intensity detection. An integrated LCD display provided the results locally. Another monitoring system for indoor environments was presented by Marques and Pitarma (2017). Their IoT solution incorporated an Arduino Uno board, a temperature and humidity sensor, a luminosity sensor and a fire sensor. Moreover, the system included an LCD for displaying information and an ethernet shield for data forwarding. Data could be stored in a database for remote access to the information. Du Plessis, Kumar, Hancke, and Silva (2016) presented another air-quality monitoring system for indoor environments, comprised of temperature, humidity, carbon monoxide and carbon dioxide sensors. Du Plessis et al. employed an ATMega88 microcontroller with a IEEE 802.15.4 transceiver for wireless communication. Results showed a 17 kbit/s effective data rate and ranges between 0°C and 90°C for temperature, with an accuracy of ±2.6%, from 45.5% to 98% for humidity with an accuracy of 3.8% and ranges from 0 ppm to 29 ppm for carbon monoxide and 0 ppm to 2000 ppm for carbon dioxide.

Finally, some studies have attempted to monitor the wellbeing of people utilising sensors. Suryadevara, Mukhopadhyay, Rayudu, and Huang (2012) presented an intelligent home monitoring system that employed data fusion in order to determine elderly people's wellness state. In order to establish this wellness state, the authors connected several sensors to home appliances and to different furniture such as beds, chairs or toilets. They presented two equations to evaluate a person's wellness based on their activity and inactivity periods. Wellness state was also considered by Lacuesta, García, García-Margariño, and Lloret (2017). The subjects' wellness state was evaluated based on heart rate variability (HRV) measured by a wristband. Users wore the wristband to visit houses they were interested in buying or renting. The system employed a point-based algorithm to determine the wellness state of the user in each home and neighbourhood and provided a recommendation based on where the user felt best.

Although there were solutions that were able to monitor indoor environments and solutions that measured certain aspects of wellness state, there weren't any solutions that incorporated both aspects, monitoring both QoL and indoor environments. In this work, we introduce QoL monitoring for indoor environments and provide a proposal on a QoL monitoring system that obtains information from the user as well as monitors the domestic environment.

QoL control and monitoring

This section presents an outline of how QoL is currently monitored. Moreover, the architecture of the system and the performance algorithm of our proposed application are presented. QoL has been measured subjectively during recent years. In fact, the WHO (World Health Organisation) defines it as "how an individual perceives their life based on the culture and the value system they live with". The individual's goals, expectations and worries are considered as well (Hubley, Russell, Palepu, & Hwang, 2014). Thus, many subjective questionnaires have been developed to evaluate both general QoL and QoL in relation to specific diseases or environments.

QoL questionnaires measure a person's satisfaction in regards to different aspects of his or her life. Aspects considered in the questionnaires are called domains. Depending on the objective of the questionnaire and the number and the themes considered in it, domains can vary. The WHO identifies six different domains on its WHOQOL-100 questionnaire (The WHOQOL Group, 1995). These are the physical, psychological and independence level, social relationships, environment and spiritual domain. Each domain is, at the same time, comprised of different facets. Facets are the activities or specific conditions from which information can be obtained in order to evaluate the domains of QoL measures. Physical health, mobility, wellbeing, fitness, energy, vitality and drug dependence are some of the facets that are related to the physical domain. Self-esteem, mental health, positive and negative feelings, stress, depression, emotions and personal fulfilment are related to the psychological domain. The independence domain may include education level, job, leisure time and activities, freedom and the physical and psychological ability to perform certain activities related to normal life, such as household chores and other everyday activities like bathing or cooking. Family life, love life, sex life, social life and all types of personal relationships as well as the feeling of belonging provided by the community are facets that can be considered for the domain on social relationships. The environment domain may include the neighbourhood, safety, domestic environment, urban environment, presence of green areas and services, transportation, privacy and other resources. Finally, the spiritual domain includes personal values and beliefs as well as religious beliefs.

These questionnaires are generally performed by the person who is getting evaluated. However, some questionnaires can be answered by a third party, such as medical staff in the event that the physical or psychical incapability of the person who is being evaluated won't allow them to complete the questionnaire on their own (Verdugo, Gómez, Arias, Navas, & Schalock, 2014). The process of evaluating quality of life can be time-consuming, as the number of questions that need to be answered may vary from 20 to 100. Nonetheless, the WHO determines facets related to mobility, work, social relationships, finances and negative feelings. Those facets allow the survey to distinguish among people with good or bad quality of life more easily (Skevington, O'Conell, & the WHOQOL Group, 2004). Therefore, shortened versions of the questionnaires can be created so as to perform a faster evaluation. Even then, there are other ways to reduce the time spent in performing these questionnaires.

New technologies introduce a new way of evaluating QoL, allowing researchers to replace questionnaires with data gathered from various sensors. New technologies allow researchers to obtain objective information on the environment and the state of a person, thus eliminating the need for collecting the same information in a subjective manner.

QoL monitoring in domestic environments

In this subsection, the architecture for QoL monitoring in domestic environments employing sensors and a QoL questionnaire implemented as an application for a smartphone is presented. The environment is one of the key aspects that influences quality of life. Unlike the rest of the domains, the environment can be improved independently of the individual. Thus, both by the will of the individual and as an initiative of the municipality, improvements in the environment and in people's quality of life can be made. In addition, not only would the quality of life related to the domain of the environment be improved, but the changes in the environment would have positive consequences in other domains. In this way, the expansion and maintenance of green areas can reduce stress levels, and play areas for children can enhance their physical and psychological levels of independence as well as the social relationship domains.

The Internet of Things (IoT) is based on providing access to the Internet to a wide variety of objects to enable communication. Moreover, wireless sensor networks (WSN) enable wireless communication among a large number of sensors. Both allow monitoring many parameters and obtaining a large amount of data that can be transported wirelessly to a database in order to be processed later. These technologies are used to monitor a large number of areas, including the environment, health, vehicles or crops. Monitoring domestic environments allows us to characterise the environment in which a person lives to determine the conditions of their home. The condition of the domestic environment can be a source of stress that decreases the quality of life of its inhabitants.

Figure 11.1 shows the monitoring system's architecture for the domestic environment and people's quality of life. On the one hand, the node collects information from the environment, measured by the sensors. After this, the data is sent wirelessly to the access point. Then, the data is sent to the database to be stored for further analysis. Concurrently, the user answers some questions about his or her quality of life through a smartphone. The data is subsequently sent to the database for analysis.

For dwelling monitoring, we propose the employment of a node and a series of sensors capable of measuring the desired parameters. Five different monitoring parameters have been considered in order to characterise the domestic environment in which the user resides. The reason these parameters have been chosen is their influence on quality of life (García, Parra, Romero, & Lloret, 2017). These parameters are temperature, humidity, air quality, brightness and noise. Extreme temperatures can affect the thermoregulatory system and cause dehydration, heat stroke or instability in movement. People with chronic respiratory diseases may also be severely affected. Humidity can enhance the effects of extreme temperatures. Furthermore, humidity can affect people with respiratory diseases and promote mould development. Air quality also has great effects on respiratory diseases. In addition, poor air quality can adversely affect quality of life and level of satisfaction with life (MacKerron & Mourato, 2009). Brightness can affect sleep

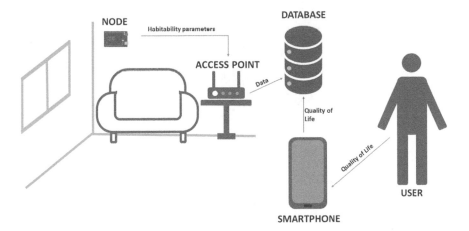

Figure 11.1 System architecture.

quality and stress levels. Moreover, brightness can affect the mood, and therefore depression levels. Finally, noise can cause problems with sleep quality and stress, as well as increasing the risk of cardiovascular problems.

Therefore, a system for monitoring the domestic environment similar to that shown in Figure 11.2 is proposed for domestic environment monitoring. In Figure 11.2, an example of a domestic environment monitoring device that performs measurements of temperature, humidity, noise and air quality is presented. The proposed system should include a sensor to measure brightness as well as the parameters monitored in the example shown in Figure 11.2. The sensors take measurements of the environment and send them to the node, which is the device that processes and forwards information. In this example, an Arduino node is being used, but the system can be developed with nodes of other brands or different features. The node processes the information and provides the wireless module. The wireless module is connected to a wireless network and sends information to the destination specified in the code. In this way, data can be collected in a database to be accessed later.

At the same time, a subjective monitoring of the quality of life of the user is proposed. In the subjective monitoring of quality of life, the user answers a reduced questionnaire. The completion of this questionnaire is intended to obtain information from other domains of quality of life in order to be able to contrast them

Figure 11.2 Domestic environment monitoring device.

with the data obtained from the sensors. The questions would be answered using a Likert scale so that the results can be statistically treated. The main domains to be considered will be the physical and psychological domains, because the studies carried out on the dwelling's parameters detect variations in people's health and mental state. Thus, facets such as fatigue throughout the day, the existence of positive or negative feelings, stress levels or physical health can be studied. Facets of other domains can also be evaluated by their relationship with the physical and psychological domain. For example, the independence level domain could be assessed by asking questions about the user's ability to carry out daily or domestic activities. These facets could be affected on days of intense heat in which people, especially children and the elderly, may suffer dizziness or heat strokes.

In order to carry out this subjective monitoring of quality of life, we propose the development of a simple mobile application that will allow subjects to answer the questionnaire easily and quickly. In addition, the information from the questionnaire can be sent to the database, thus obtaining the information in a considerably shorter time than performing this test on paper. Figure 11.3 shows an example of the application display for QoL monitoring using the Likert scale. In this display, the question about the facet that is being evaluated appears at the top of the screen. The user must press the description corresponding to his state. Different emoticons, colours and texts have been assigned to each of them, allowing users to quickly understand the evaluation process. As a result, users perform this test more

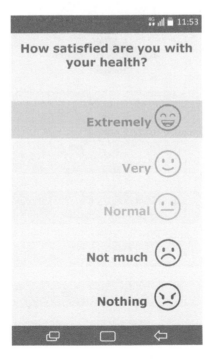

Figure 11.3 Display of a question during the subjective evaluation of QoL employing a smartphone application.

pleasantly allowing them to distinguish the available answers in an easy manner. In addition, the display's large size is user-friendly for those with visual difficulties. Once all the questions have been answered, the information obtained will be sent to the database, where each user can be linked to their questionnaires using an ID.

Some aspects to consider may be the request for another type of information that allows researchers to assess the obtained results according to the area or people's characteristics. Thus, before the users begin the evaluation of their QoL, they can be asked to indicate the neighbourhood in which they live, their sex or their age range. In this way, the results of QoL for different ages can be compared and the differences perceived by men or women in QoL can be assessed.

Figure 11.4 presents the performance algorithm of the application. Each one of the questions of the WHOQOL-BREF (Yao, Chung, Yu, & Wang, 2002) test must be answered by the user. The application does not allow the user to skip a question, so all performed tests are considered valid. When the 26 questions are answered, the mean value for each domain is obtained. In this case, four domains

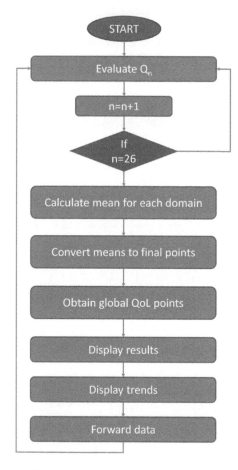

Figure 11.4 Algorithm of the proposed app.

are evaluated: physical and psychological health, social relationships and environment. Then, the obtained values are numbered again according to a 10-based scale using the typical point system utilised in Spain. Finally, the global QoL punctuation is calculated and the results are provided to the user. The trend is also calculated in order to assess the evolution of the user's QoL.

The performance of QoL assessments and housing monitoring can result in a large amount of data. To deal with the data, a series of statistical tests is proposed. The data obtained from the sensors have five variables. The QoL of the inhabitants of the dwelling can be treated as a single result obtained from the average of each domain, or it can be separated according to the different evaluated domains. In this way, a greater number of variables are obtained. By performing a multivariate statistical analysis, the variations in the environment will be analysed to determine if they have produced any variations in the person's QoL. Each variable of quality of life will be compared to the results obtained from monitoring the dwelling. Thus, variations in people's quality of life can be identified. Whether or not the causes of these variations are related to the environment in which people live can also be determined in this process. Thanks to this development, a solution that improves residents' quality of life can be proposed.

Results

This section presents the measurements of some parameters typically assessed in home environments. The selected home is the fifth floor of a building located on the Mediterranean coast. This zone is characterised by presenting very high levels of relative humidity due to the closeness of the sea. The house has a living room, a kitchen, a bathroom and two bedrooms. The distribution of this dwelling is shown in Figure 11.5. The measurements are performed in the biggest room. The node is located

Figure 11.5 Home distribution.

in the centre of the room at 1 m high with respect to the ground. Measurements were performed for 19 hours where the parameters varied according to the time.

Of the measurable parameters in a house, temperature and relative humidity are the easiest to measure and have a greater influence on people's welfare. Figure 11.6 shows the temperature (in °C) and relative humidity (in %) measured in the room under study. One of the important aspects to be observed is that temperature and relative humidity increase proportionally. The lowest temperature value is registered in the central hours of dawn.

The level of light is important to develop daily activities at home, for example, low light levels are important during sleep hours. Figure 11.7 shows the darkness

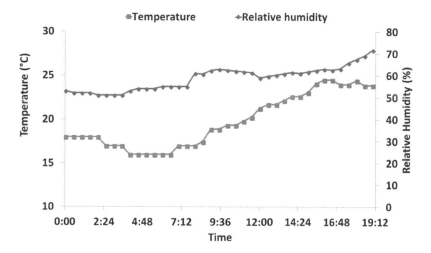

Figure 11.6 Temperature and relative humidity.

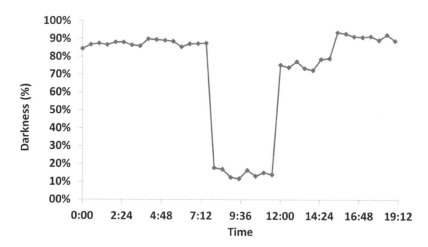

Figure 11.7 Darkness levels.

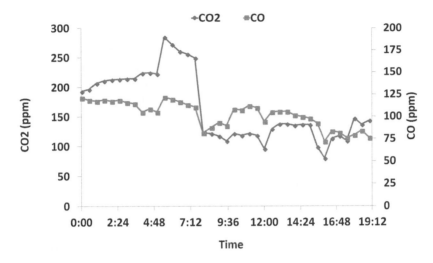

Figure 11.8 Values of CO and CO_2 to measure air quality.

levels measured in the room, where 100% represents total darkness. As we can see, the darkness levels are high enough, except for the period between 7:20 and 12:00 where the blinds have been opened to ventilate the room and the sunlight directly enters through the window.

Finally, Figure 11.8 shows the values of CO and CO_2 measured in the room. Because these values usually remain quite stable in an inhabited house (a normal situation is where there are no sources that consume oxygen from the air, such as a gas stove), we have released CO_2 in the room (levels below the danger limits) and after that, we have opened the window. As we can see, before opening the window, we reach values of 285 ppm and after opening the window, the CO_2 value decreases up to 125 ppm. The value of CO also decreases when the window is opened.

As we have observed, the measured parameters are within the limits established as normal for living in a house. It would be possible to design an intelligent system for domestic environments by using the correct programming to establish security thresholds. It is then possible to correlate these values with others such as possible symptoms of diseases for elderly and disabled people. With this information, the system could notify close relatives or even activate domotic systems to provide a solution for an abnormal situation.

Conclusion and future work

QoL monitoring is important because it is the way in which people perceive their degree of satisfaction with the various aspects of their lives. New technologies help to facilitate this monitoring, thus reducing the timing in which it is carried out. The environment is one of the different domains that influence QoL. In this chapter, we propose the implementation of a housing monitoring system that uses various sensors to obtain information on temperature, humidity, luminosity, air quality and noise in the domestic environment. At the same time, we propose the development of a simple mobile application that allows the subjective assessment of other domains of quality of life. This simple mobile application will enable researchers to study the effects of the environment on the quality of life perceived by the user.

As future work, we will employ sensors to monitor the parameters of people living in domestic environments, such as heart rate and body temperature. This way, QoL results can be compared to people's vitals and the state of their environment and more specific correlations can be found.

References

Bedi, G., Kunar, G., & Singh, R. (2016). *Internet of Things (IoT) sensors for smart home electric energy usage management*. IEEE International Conference on Information and Automation for Sustainability, Galle.

Cavanaugh, J. T., Coleman, K. L., Gaines, J. M., Laing, L., & Morey, M. C. (2007). Using step activity monitoring to characterize ambulatory activity in community-dwelling older adults. *Journal of the American Geriatrics Society, 55*(1), 120–124.

Chen, S., Chang, S., & Chen, Y., (2015). Development of a multisensor embedded intelligent home environment monitoring system based on digital signal processor and wi-fi. *International Journal of Distributed Sensor Networks, 11*(6), 171365.

Coni, A., Mellone, S., Leach, J. M., Colpo, M., Bandinelli, S., & Chiari, L. (2016). *Association between smartphone-based activity monitoring and traditional clinical assessment tools in community-dwelling older people*. IEEE 2nd International Forum on Research and Technologies for Society and Industry Leveraging a better tomorrow, Bologna.

Du Plessis, R., Kumar, A., Hancke, G. P., & Silva, B. J. (2016). *A wireless system for indoor air quality monitoring*. 42nd Annual Conference of the IEEE in Industrial Electronics Society, Florence.

García, L., Parra, L., Romero, O., & Lloret, J. (2017). System for monitoring the wellness state of people in domestic environments employing emoticon-based HCI. *Journal of Supercomputing*, 1–25.

Hubley, A. M., Russell, L. B., Palepu, A., & Hwang, S. W. (2014). Subjective quality of life among individuals who are homeless: A review of current knowledge. *Social Indicators Research, 115*(1), 509–524.

Lacuesta, R., García, L., García-Margariño, I., & Lloret, J. (2017). System to recommend the best place to live based on wellness state of the user employing the heart rate variability. *IEEE Access, 5*, 10594–10604.

Leidinger, M., Reimringer, W., Alépée, C., Rieger, M., Sauerwald, T., Conrad, T., & Schütze, A. (2016). *Gas measurement system for indoor air quality monitoring using an integrated pre-concentrator gas sensor system*. Proceedings of the GMM-Fb. 86, Mikro-Nano-Integration, Duisburg.

Lita, I., Visan, D. A., Cioc, I. B., Mazare, A. G., & Teodorescu, R. M. (2016). *Indoor environmental parameters monitoring for building automation systems*. International Conference on Electronics, Computers and Artificial Intelligence, Ploiesti.

MacKerron, G., & Mourato, S., (2009). Life satisfaction and air quality in London. *Science Direct*, *68*(5), 1441–1453.

Marques, G., & Pitarma, R. (2017). *Monitorizaçao e Controlo do Ambiente Interior*. IEEE 12th Iberian Conference on Information Systems and Technologies, Lisbon.

Skevington, S. M., O'Conell, K. A., & the WHOQOL Group. (2004). Can we identify the poorest quality of life? Assessing the importance of quality of life using the WHOQOL-100. *Quality of Life Research*, *13*, 23–34.

Suryadevara, N. K., Mukhopadhyay, S. C., Rayudu, R. K., & Huang, Y. M. (2012, May 13–16). *Sensor data fusion to determine wellness of an elderly intelligent home monitoring environment*. IEEE International Conference on Instrumentation and Measurement Technology, Graz.

Verdugo, R. A., Gómez, L. E., Arias, B., Navas, P., & Schalock, R. L. (2014). Measuring quality of life in people with intellectual and multiple disabilities: Validation of the San Martín scale. *Research in Developmental Disabilities*, *35*(1), 75–86.

The WHOQOL Group. (1995). The world health organization quality of life assessment (WHOQOL): Position paper from the world health organization. *Social Science & Medicine*, *41*(10), 1403–1409.

Yao, G., Chung, C. W., Yu, C. F., & Wang, J. D. (2002). Development and verification of the validity and reliability of the WHOQOL-BREF Taiwan version. *Journal of the Formosan Medical Association*, *101*(5), 342–351.

12 Landmark preservation criteria and BIM methodology

Giacomo Sorino, Antonio Jiménez-Delgado, Carlo Manfredi

Introduction

Among the several architectural practices applied to a built heritage, the conservation of the latter is the one that mostly involves the sensibility towards various environmental issues related to the materials, the life cycle and, last but not least, the overall functional identity. In recent times, the Mediterranean area has experienced an unrivalled alteration regarding the demographic articulation compared to similar occurrences in European countries. The Mediterranean area is experiencing social and economic growth that is comparable to the growth that occurred in other countries during the phenomenon of industrialisation in previous centuries. What was recorded during the last few years reveals a significant exploitation of natural resources. The immense and, according to several surveys, uncontrolled growth of the urban settlements close to the Mediterranean coast, which used to be agricultural territories, is related particularly to the spreading of different, in the context of tourism, residential and commercial functions which provoked a settling sprawl. Such sprawls are expressed with different connotations defined by phenomena of land consumption and caused a large part of the original settlement characteristics to be confused with a new and unrecognisable form of landscape.

In this analysis, it is evident that the built heritage, although reduced in its consistent character and being deficient, compared to the indistinct mass of more recent constructions, raises an exceptional interest due to its connection with the past and the direct relation between the historical settlements within the structure of the territory. The survival of those stratified building-type characters is a foothold to which to anchor the identity of contemporaneity, and in which to rediscover the reasons for a structure of meaning, even before a functional one. Knowledge and reuse of historical heritage can guide choices and behaviours in light of a deeper understanding of the meaning of presence on the territory.

It is an epochal change, not yet completed, but which has led to the deepest mutations we can read in the face of the landscape. In addition, the historical centres were in turn made objects of reuse for purely touristic purposes, with short-term rentals of extended portions. These historical centres are real pieces of the city that have experienced an impoverishment of the social fabric as well as the expulsion of the poorer classes and the bleakest gentrification.

Despite the loss of large pieces of urban fabric, now devoted to an economic return that can be quantified in detail in the costs "per room per night", the identity matrix of the ancient settlement continues to be a reference criterion to keep in mind in light of future developments. The age-old relationships that have sedimented a certain urban form have not lost their ability to function and charm in qualitative terms regarding the possibilities that the future city can offer.

All these changes are taking place at a time when the gradual tendency towards the reappropriation of historical centres, which has been going on for over the last 50 years in the most economically advantaged areas, have opposed the degradation of large central parts of the ancient city. The reappropriation of the historical centres of these inner city parts has completely reversed the tendency to search for favourable areas outside mediaeval settlements. These mediaeval centres have consistently survived, almost unchanged, throughout the centuries until the 19th century (and beyond, as the example of the last *îlots insalubres* of a city like Paris).

As almost invariably happened in the affair of the Anthropocene, this profound mutation has been accompanied by a technological development that has sustained it. Technological changes in the approach of architectural issues have also been observed. In view of fast-growing technologies supported by the giants of the technological monopoly, which is witnessing critics as well as supporters, and that is already pervading our lives, building information modelling (BIM) promises to unify the project and the realisation of a building in a single process that links the control of the detail to the detailed definition of the materials.

It is evident, however, that doubts remain regarding the use of these instruments in application to the built heritage. The difficulties of knowing an existing building are connatural to its condition of consistency. The condition of consistency remains partly unknown due to the difficulties of recognition that can often be incomplete. In addition, the loss of continuity from the ancient building ways of production are only in part now repeatable. It is therefore evident that there are still hesitations in adopting new approaches such as BIM, both due to the lack of skills extended to large areas and to the evident difficulty of abandoning well-established and well-known design and construction processes.

The BIM methodology and its application in the preservation of built heritage

What is BIM? And how can this method be applied to the field of restoration of built heritage while considering it as a new frontier on the existing conservation criteria?

> BIM is a digital representation of the physical and functional characteristics of a building. As such, it serves as a shared knowledge resource for information about a building, forming a reliable basis for decisions made during its life cycle from inception onward. The key principal is that BIM is not any single act or process. It is not creating a 3D model in isolation from others

or utilising computer-based fabrication. It is being aware of the information needs of others as you undertake your part of the process. A BIM model can contain information/data on design, construction, logistics, operation, maintenance, budgets, schedules and much more. The information contained within BIM enables richer analysis than traditional processes. Information created in one phase can be passed to the next phase for further development and reuse

(New Zealand, 2014, p. 4)

Building Information Modeling (BIM) is one of the management technologies that is causing the most debate architecture, engineering and construction fields (Architecture, Engineering and Construction, AEC). BIM technology allows building digitally accurate virtual models of a building to support all phases of the building process, thus allowing more efficient analysis and control than traditional processes. Once these models have been completed, they accurately contain the geometry and data necessary for several successful steps: the design phases, the choice of contractor, the construction and subsequent management of the useful life of the building.[1]

(Eastman, Teicholz, Sacks, & Liston, 2016, ch. 1)

It emerges from these statements that the paradigm shift does not concern only the graphic representation of the architectural project which focuses on the three-dimensional component instead of the two-dimensional component, but on the contrary, aims at renewing the whole structure and the entire building process from acquisition of data up to the implementation and management of the building.

At any time during the building process the designer must necessarily reason in a complex way, considering both the phases prior to his work with knowledge and planning and the subsequent ones, those of realisation. BIM is proposed as an instrument for complex areas such as architectural design, plant engineering, structural, infrastructure, site management, facility management and project management, off-site machining, prefabrication (although its use is extremely limited in specific conservation projects), assembly, planned maintenance and even building restoration.

The ultimate goal of BIM is therefore to be able to control and manage through a single information model all phases of a building's life cycle, thus minimising the possibility of error between the various phases and between the transfer of skills between various actors in the building process. For this to happen, a complete integration of all the areas concerning an architectural building is necessary. This integration is needed in order to organise the entire process of design, development and management through "facilities", i.e., features and services that act as starting points.

A building has several phases: projecting, building, existence also known as "useful life" and dismissal or recuperation. The "life phase" of the building, which is the most durable phase in terms of time, depends mainly on the materials with which the building was designed and on the management of their characteristics over time. With facility management, the aim is to optimise the available resources both during the project phase and during the next management phase, carefully planning the ordinary and extraordinary maintenance and, in its last landing, the future reconversion or demolition of the building at the end of its useful life.

Such a process, regardless of the application or the lack thereof of BIM, can benefit from a correct and functional representation of the built heritage. This process can also benefit from a correct integration between the different skills and in retrieving and analysing information that concerns the various elements in all their respective aspects. Therefore, having a virtual prototype that is being continuously updated would aid in extracting historical information, assets and techniques. Having the virtual prototype would be of great help, if not essential, however, for a proper management of the "facilities" and therefore to correctly plan the resources and operations necessary for the conservation of the built heritage over time.

As shown later, we begin to think about a digitalisation of existing built heritage, thus creating a multimedia catalogue accessible to all users of the AEC sector (architecture, engineering and construction). For easy management of information, digitalisation of existing built heritage of the intervention happens through using the parameterisation of data possible through the BIM platforms. In the case of historical built heritage that requires restoration, however, the sole process of digitisation for the building is the chosen one for an easy and quick planning.

There is a substantial delay in the Mediterranean countries in the application of a methodology that is now widespread in Northern Europe and in non-European countries. This reluctance of new methodological application shows the difficulty of intervening on a heritage that is perhaps not larger but certainly with higher complexity. The delay in Mediterranean countries in the application of newer methodology is also due to resistance shown by the building sector and legislation in adopting more advanced technological paradigms.

BIM applied to existing buildings

Within the scope of restoration and conservation, we are increasingly appreciating, even if through a gradual process, the benefits that this new paradigm can bring in relief and diagnostics. In the last decade, thanks to the greater economy of technologies such as laser scanners and increasingly accurate photogrammetric returns through high-quality cameras, BIM applied to restoration projects has been developed in the investigative field.

HBIM (historic building information modelling) consists in the return of geometrical and descriptive digital models of historical buildings, based on new innovative survey techniques, to which a whole series of parametric information can be connected that describes, not only graphically, but also characteristically, the building properties and each of its individual components. The methodological process applied by HBIM is not completely different from a traditional process in the restoration field; rather, they correspond in almost all the steps. The substantial difference consists of the tools adopted by the architect during the various phases and in how the results produced are subsequently collimated in a single complete informative database. Moreover, in this particular historical period of the last decade, it is not possible to disregard traditional knowledge and methodology, since it is not yet ready for a complete transition to the BIM methodology, even less in Mediterranean countries.

A gradual transition from the traditional methodology to the BIM method takes place through a hybrid workflow, theorised by Grabowski (2010), according to which the transition from AUTOCAD (Autodesk program based on Computer-Aided Drafting methodology) to BIM must be gradual and supported by different approaches so as to allow a gradual but competitive insertion within the consolidated current methodology.

> BIM software is becoming common-place, yet design software that uses 2D linework will remain popular for the foreseeable future. These two different approaches to design mean that a hybrid workflow becomes a part of everyday practice. For design firms to remain competitive, it is worthwhile for them to establish a solid hybrid workflow – for themselves, their partners, and their clients.
>
> (Grabowski, 2010, p. 32)

By dividing the phases of the HBIM process into three macro phases, the output data of each phase corresponds to the same time of the input data for the next phase. Data collection must take place through an image-based or range-based photogrammetric approach realised respectively through a cheaper photographic approach or a more expensive but far more accurate approach through laser scans. The correct execution of an HBIM survey can also include the combination of the last two methodologies, and in some cases it can also include the support of a traditional survey.

The processing and collation of the data collected and obtained by various software through "point cloud" therefore involves the generation of an informative parametric model within which all information concerning the historical building (geometric, historical) is inserted directly and indirectly. The information is then extrapolated through schedules, graphic tables, libraries of parametric objects, energy information, etc.

All this information allows cataloguing and analysing every single component modelled according to its real characteristics, thus producing an "as-is" model, which, in the case of built heritage, translates into an "as-damaged" model, paying particular attention to the "I" of BIM; that is, to the pathological diagnostic *information* that characterises this element in a precise historical moment.

The understanding of a built heritage through its digital return

Understanding an architectural building does not always coincide with designing an architectural building. Understanding an architectural building involves a study, a deepening, a greater effort from the mere graphic or metric rendering of an object.

An effective understanding is fundamental nowadays in the restitution of an architectural artefact, since without the subject's understanding of it, there would not be correct planning, correct analysis or correct valorisation of the heritage.

In order to really appreciate the innovative process of HBIM, it is necessary to develop the model starting from "remotely sensed" data, i.e., a relief realised with techniques and technologies that do not require physical contact with the historical heritage, often ruined by human presence or simply inaccessible. A correct geometric rendering of a historical heritage is not enough to describe every aspect of it, nor is it sufficient to undertake a restoration project that touches on every relevant architectural aspect. Instead, it is necessary to obtain mainly information that can return, even if at a later time, visual or theoretical data on the artefact that a simple traditional relief could easily overlook. It is also advisable to retrieve archival historical data concerning the structure or stratification of the walls so as to be able to reconstruct the different modifications that took place during the decades that are supported by field surveys.

As already mentioned, thanks to technological innovation in the photographic field and to the development of more advanced personal computers, in recent years, the theme of photogrammetry has been developed as a support to architecture, thus succeeding in integrating geometric modelling tools with instruments of parametric modelling. The well-established architectural survey campaign is therefore destined to change over time, increasingly becoming a remote data collection campaign rather than a real intrusion into the building. In many cases, it is dangerous to create reliefs due to a property's degradation, and in other cases it is equally cramped and uncomfortable where there are semi-inaccessible places or with little possibility of manoeuvring. The best option to bypass these kind of problems is to use a 3D laser scanner, which can provide a photogrammetric digital model based on point cloud. That photogrammetric model can be used as a data input into the parametric model due to its efficiency to return all the geometric information up to the thickness of walls and interspaces.

A further feature of the point cloud/photogrammetric model is to be based on photograms that overlap with certain geometric-digital rules, giving the model the ability to extract texturised photorealistic orthophotos, and thus deriving photogrammetric model data directly from the input. The restitution therefore, although not parametric, is mostly complete from the geometric point of view and the yield of the materials, guaranteeing the designer continuous consultation of the model in place of a further site survey.

There are four obvious advantages of 3D laser scanner relief technique or photogrammetric image-based technique compared to traditional survey techniques:

- High-detail restitution in situations of particular complexity
- Reduction of relief and restitution times
- Acquisition of simultaneously 3D point clouds and colour photos
- Integration with 3D CAD and BIM modelling systems

The advantages of using a photogrammetric model or a laser scanner are therefore numerous, both for the present building study and for a future usability of certain information. The flexibility of the applied technology and the potential that said technology has, if integrated with drones and topographic instruments, turns out to be both a point of arrival for the new restoration project and a starting point for innovative management and knowledge of cultural built heritage, thus restoring importance to the historicity of the architectural heritage. This model is as detailed as it is effective and allows the modeller to complete the cognitive framework that a built heritage restoration intervention requires; that is, a pathological diagnostic study.

Traditionally, we can help in the identification of a degradation-process assessment through the photos taken during the campaign for relief recognition. This campaign for relief recognition utilises the above-mentioned edited photos or more or less invasive investigations correlated with notes and diagnostic studies. The aim of HBIM, in addition to graphically mapping the degradation in order to return graphical tables on the surfaces that present pathologies, is to be able to reconstruct a virtual prototype that possesses the complete pathological study in its database form, thus providing an understanding of the historical and archive data on the building. The materiality, the geometry, the pathological picture, the architectural composition and subsequently the historical and archival information form what is considered the final output data of an HBIM project, which is the parametric model

Several factors aid in obtaining a prototype: various software that doesn't belong to BIM platforms such as Rhinoceros and Photoscan, the integration of diversified skills and the integration of innovative technologies, methodologies and traditional information as well as attitudes. Through a combination of these factors, we obtain an "intelligent" and parametric virtual prototype that can return every single piece of information inserted inside it through graphic tables, schedules and parametric representations that allow an optimal manoeuvrability of the instrument and an easy interpretation of the output data.

As is often the case for architectural modelling of a virtual prototype, it is especially important, at least in this phase of BIM implementation, to cope with scarce resources or logistical and cultural difficulties. The integration between traditional methodology and innovative methodology according to the hybrid workflow theory is an important one, provided that the former is placed at the service of the second.

Reverse engineering, photogrammetry and parametric modelling

To better understand the change in paradigm that this new BIM method involves, we need to introduce some issues related to reverse engineering when it is applied to the field of architecture. Usually, during everyday life, we think in two main ways in order to obtain a complete reasoning: deduction or induction. According to Ch. Peirce, on the other hand, the only way to reach an increase in knowledge is through abduction, the condition underlying reverse engineering:

> Abduction is the process of forming explanatory hypotheses. It is the only logical operation that introduces a new idea, since induction merely determines

a value and deduction simply develops the necessary consequences of a pure hypothesis. The deduction finds that something must be; induction shows that something is really operative; abduction merely suggests that something can be.

(Peirce, 1935).

3D scans or photogrammetric restitution are the means through which the restoration project and the BIM project can apply a form of abduction. This form of abduction starts from a real photogrammetric object reproduced in the form of a point cloud to reach a CAD or BIM model parametric, which fully describes the characteristics while being a virtual prototype. Thanks to stereo matching algorithms and to instruments that promote the idea of reverse engineering, we arrive at the HBIM survey of the actual state of a historic building.

As customary, and as it should still currently be, the survey phase of a built heritage is considered the cornerstone of a good restoration or conservation project. The ability of an architect to appreciate the work to which he or she is going to relate, the attention and accuracy with which a technician performs analysis on materials or structural analysis on the building, the careful attention of the operators to the individual details effectively captured with drawings and photos, are all qualities necessary to start a good campaign that responds successfully to reality in its digital transposition.

As exemplified in the decades of Brunelleschi or Vasari, it is good not to lose the ability to relate to the architect's heritage, in a sort of poetic physical and spiritual contact: "The architect's own mental attitude and a particular condition of the spirit that expresses itself in a desire for knowledge and emotional involvement during the perception and analysis of the built space" (Jaff, 2005).[2] In addition to these characteristics, a survey in the BIM field requires technological skills for remotely sensed data collection and for the transposition of said data into a virtual prototype. The connection between the architect's emotion and the physical and emotional detachment given by drones, laser scanners and photogrammetry produce what is actually a point of arrival in the acquisition of data, but also a starting point in the BIM project. "In general, photogrammetry is defined as the set of processes for the use of photographic images for the formation of topographic maps and for the execution of architectural surveys" (Jaff, 2005).[3]

Indirect relief is the method of obtaining photos of a built heritage through photography or scanning. There are two types of indirect relief: range based, which operates through the emission of an electromagnetic signal recorded by the instrument (laser scanner) in order to derive characteristics and distances of the artefact by developing a cloud of points, and image-based, which operates through passive sensors and light for the acquisition of images that will only be reprocessed later.

The range-based method of data acquisition is certainly more detailed and effective, above all because it is based on electromagnetic signals and it allows one to obtain the thicknesses of architectural elements with relatively negligible errors. However, since it is still a rather expensive technique and a campaign to raise awareness on the methodology in the Mediterranean countries has never been launched, it is advisable to investigate, at least in this area, the second technique of restitution, photogrammetry. Two types of photogrammetry are distinguished, namely aerial photogrammetry, also known as "the distant ones" executed with drones, and the terrestrial photogrammetry, also called "of the neighbors".

Definitely cheaper and certainly more flexible in terms of transportation and availability of the material, photogrammetry in general provides an excellent contribution to the study of the building, despite not being able to detect hidden thicknesses. The integration of the two techniques, with the help of mobile supports (tripods) for the terrestrial method and remotely controlled aerial drones for the aerial one (topography), can provide a database rich enough to describe the entire building complex from the outside. The goal is therefore to create a BIM model starting from a hybrid database, which contains photogrammetric information and which is as economical and flexible as possible so not to allocate significant economic, instrumental and logistical resources.

Using SfM (structure from motion) algorithms, completed by stereo matching algorithms, the post-production of the base frames becomes of fundamental importance for the success of a terrestrial photogrammetric model. As seen in the human eye, photogrammetry is based on principles of stereoscopy; this also allows one to obtain the depth of objects by shooting different frames with different angles and different grip points. The operational difficulty consists precisely in realising a vision that is as close as possible to the real one. To do this, shots that allow an overlapping of 60–70% of the surface must be taken so as to cover the entire surface under examination, but at the same time obtain a large number of overlapping points and comparable geometries for data processing (Monti & Selvini, 2015).

Secondly, it is appropriate to adjust both the focal distance by keeping it fixed during the survey phase as well as the white balance in order to obtain images that can be interpolated later, when the digital restitution through programmes like Agisoft PhotoScan and Autodesk ReCap is switched. From an organisational and timing point of view, the remotely sensed technological survey is therefore less complex and requires a much shorter timeframe and workforce, even if it is more specialised.

In recent times, the condition under which this step can take place was the advent of SfM, which drastically reduced the imbalances between acquisition time and return time of the three-dimensional model that characterised the first photogrammetry, as well as the non-transportability and availability instrumentation, such as digital metric rooms and professional computers, which have been replaced by reflex cameras, laser scanners and common laptops. Any professional can therefore apply these techniques to any type of project and with few resources (the only necessities are an operator with a digital camera and a laptop), thus ensuring more safety in relation to hazardous buildings and decreasing the time of the survey itself.

On the basis of this data, a new methodological paradigm is created which flows into the parametric modelling of the building and incorporates any type of characteristic possessed. The ability that this new model will have to fully describe the heritage in question and to be able to carry out a series of studies aimed at preserving it, is what distinguishes it from a simple 3D model that until now had the purpose of explaining the building geometrically.

It is impossible to find a wide range of information belonging to the traditional survey (cadastral data, structural data, detailed measures of the elements). However, at the same time it will be possible to develop an interdisciplinary and above all constantly updatable project, which will host all this information in a diversified way. Moreover, especially in the HBIM area, the particular modelling of individual architectural details, treated as separate elements and subsequently assembled, turns into a great advantage at the time of the pathological diagnostic study of the building, both from a graphic point of view (mapping the degradation) and from an information point of view (internal database to the information component) (Di Giuda, 2017).

It is in fact known that a single column of the Parthenon or the Pantheon possesses often different characteristics, and indeed, often every single piece of stone that makes up the column presents a unique material, composition, pathology and state of degradation. In the same way, both in a restoration project and in the survey of a built heritage, every single element is rightly treated separately, entering in detail as much as possible to return a level of development (LOD) congruent to a built heritage.[4]

Heritage management through BIM

How can parametric modelling be useful for a restoration project or for integrated building management? There are various forms in which a conservative study of an architecture can be carried out. Certainly, each of them leads to the result of an improvement in physical and structural conditions; nevertheless, the possibility of being able to carry out the study within a single "home", and therefore to be able to extrapolate information often related to one another by a single tool, is an opportunity that a BIM designer can take into account.

The ability to be able to consider the model as a set of primary models, which make up each real element, makes it possible to carry out both a general and a particular study of the elements. This study is not only geometric and volumetric, but above all material, pathological and integrated with all the disciplines of architecture. Besides creating the classic relief of the various facades' surface pathologies, parameterising the model through shared parameters and design parameters can obtain infinite information about the element. The information derived from the parameterisation of the model is subsequently extractable in tables or abacuses, thus ensuring a database not only through graphs explaining the state of degradation but also through an informative database consisting of the diagnostic framework.

It is the ability to extract information of any kind and therefore to be able to collaborate with various areas that allows one to take advantage of project management and development. The possibility of interacting with different actors in the building process is in fact essential if you want to involve any designer and any aspect that concerns a built heritage. To do this it is necessary to implement an interoperable process on two main lines: the first concerns the possibility to interchange data between different software and manufacturers through different formats and through Industry Foundation Classes (IFC);[5] the second concerns communication between the actors of the building process, which will necessarily have different skills and the ability to use software that is certainly diversified. In order to allow a real global use of the virtual prototype, the management of the building itself therefore becomes in some aspects more important than the restoration project itself.

The simple diagnostic-pathological comparison over time, the possibility of managing different project phases or different construction phases that have developed in the past, or even the extraction of parametrised tables and schedules in common formats such as Excel, guarantee that the HBIM project is interoperable and integrated in a collaborative capacity, thus favouring easy management of the building.

The process of remote viewing and virtual reality are even more integrable and interoperable. In this particular historical period it is necessary that any collaborator in an architectural project can access any information considered important in order for the work to develop. The prerogative of the knowledge of BIM programmes and platforms can therefore be a collaborative obstacle where the skills are not homogeneous between the different actors. This is why we use different viewing platforms (Tekla BIMsight and Solibri Model Viewer) that allow us to expand the range of action of the virtual prototype, reaching even people who do not have knowledge in the BIM field.

Even more effective is the application of virtual reality (VR). Originating from a "gaming" environment, the visualisation of a "full immersion" model serves to guarantee a wider experience of the model and can be of three types: virtual reality, augmented reality and mixed reality. In the phase of restoration or management of a built heritage, the VR objective works to ensure a three-dimensional experience. The opportunity that virtual reality provides during a project is to elicit a certain type of depth of feeling and visual. This depth of visual gives the opportunity to interpret and understand the architectural heritage even to users who are not in the sector of buildings and architecture, such as tourists or common customers.

Technologies like wired gloves (gloves), visors and helmets for the detection of movements can be, in the near future, the new frontier for communication between collaborators or between client and designer, thus making the understanding of the project status "as damaged" or "as is" much simpler and more immediate and therefore avoiding misunderstandings and delays in timing. Both collaboration and interoperability as well as optimal communication are fundamental and indispensable architectural components, especially in projects where the understanding of architectural heritage is difficult but at the same time fundamental.

Case studies

The ability to apply such methodologies to any case of conservative restoration study constitutes an added value to technological support; it is no longer understood as something "expensive" or reserved for a few interlocutors. In extreme cases, for example where advanced technological support is still a mere future hypothesis, the application of the methodologies presented is not taken into consideration; the transportability and the economic efficiency of the instrumentation used predominate, as well as the speed of execution of the relief phase.

A clear example of how this methodology can be applied with the minimum of the necessary resources can be drawn from the thesis work carried out in the city of Tetuán (Morocco) for the Polytechnic of Bari entitled "BIM Methodology and Relief for Traditional Architectures". The goal of the work, contrary to what is usually achieved trying to take advantage of the most advanced and expensive technologies available for demonstrating a certain paradigm, was to realise a photogrammetric survey and a subsequent parametric modelling with the minimum possible resources, even if this sometimes involved uncertainties and incompleteness of the result. With a common professional Reflex camera and an adjustable tripod, several typical Moroccan "house-patios" have been investigated and one case study was chosen (see Figures 12.1, 12.2 and 12.3).

It is evident that in many cases, not having the appropriate equipment (such as laser scanners in certain architectures) and having chosen an architectural typology with narrow and often not very bright spaces, photogrammetry has sometimes proven ineffective for correct three-dimensional modelling. Nevertheless, a preliminary photogrammetric study could be carried out on different Moroccan house-patio. This preliminary photogrammetric study should positively bring to light the extent to which a photogrammetric survey can be pushed without losing quality in the result, thus allowing a subsequent processing in BIM.

Figure 12.1 Tetuán, Building 1. Photogrammetry.

Figure 12.2 Tetuán, Building 2. Photogrammetry.

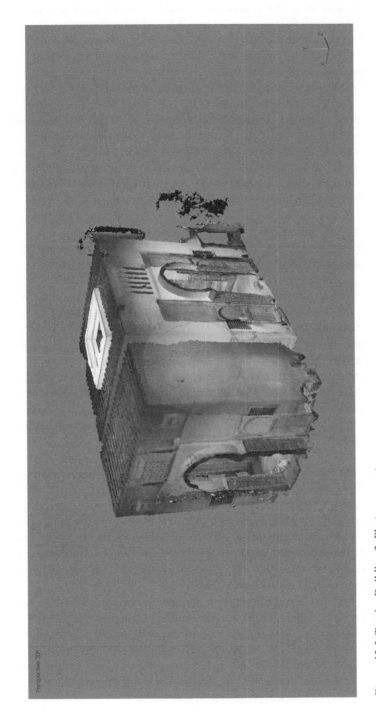

Perspective 33°

Figure 12.3 Tetuán, Building 3. Photogrammetry.

During the case study, it was therefore possible to demonstrate how any operator in the architectural field with basic knowledge of photogrammetry and photography of the survey methodological study could make a successful digital survey, regardless of the adversity of the site and the technological lack. One of the case studies proved to be suitable for digital representation through Agisoft Photoscan and Autodesk ReCap, producing a photo model of the building's internal patio characteristics.

As explained in the previous paragraphs, the hybrid workflow is necessary for the success of a complete study, and even more so in this case where not all information can be extrapolated from a photogrammetric survey. The application of the hybrid workflow methodology does not exclude and does not diminish the importance of the photogrammetric survey, which is appropriately integrated into the BIM platforms but is nevertheless a starting point and an added value for the understanding and for the realisation of the virtual prototype through the cloud point export (see Figure 12.4).

The next parametric modelling (see Figures 12.5 and 12.6) however, has been greatly facilitated by the photogrammetric survey for several reasons:

- the possibility of extrapolating photographic, volumetric and pattern information that allows us, at any time and in any place, a subsequent realistic consultation of the case study;
- the possibility of obtaining different and often more accurate measurements compared to traditional measures; and
- pathological diagnostic analysis is no longer only in the form of surface relief pathologies of the various facades or an information catalogue. In addition, information integrated into the single object modelled in the virtual prototype can be utilised in the use of BIM, thus helping form a unique digital process (see Figures 12.7 and 12.8).

The basis of this study requires a cataloguing of the possible pathologies present on the built heritage so that a methodological guideline can be easily applied to other case studies and architectures, thus maintaining the same approach in the different restoration hypotheses. In fact, in the near or remote future, the possibility of being able to make a catalogue of the *viviendas* of the Medina of Tetuán is very stimulating both from a historical point of view, allowing us to store a realistic and three-dimensional photo of a historical centre for UNESCO, but also from an engineering point of view (Líndez, 2014).

The comparability of these descriptive parameters realised on the basis of pathological diagnostic manuals and the continuous parametric updatability within the model, allows for a 360-degree study of the built heritage and the comparable other built heritage, both during the different phases of the study but also in completely isolated periods of time. The management of the virtual prototype is therefore strictly related to the possible management of the building in every aspect and in every historical phase (see Figure 12.9).

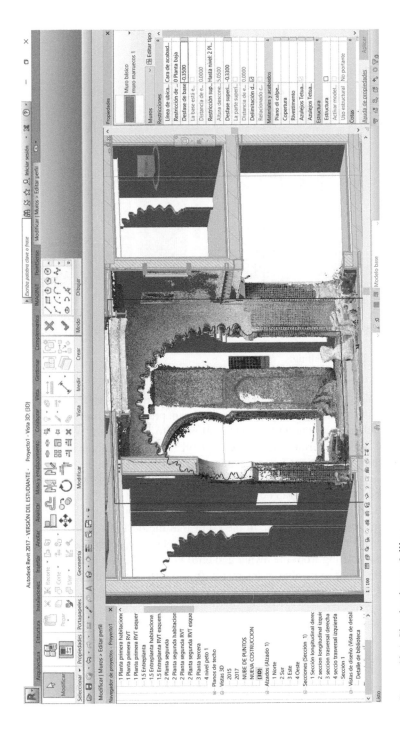

Figure 12.4 Cloud point modelling.

Figure 12.5 BIM modelling.

The possibility of being able to manage different design phases, combined with the different historical periods as well as different project proposals in turn related to the actual status of the built heritage, allow us to examine the whole historical process of the built heritage itself (see Figure 12.10). The ultimate goal is to safeguard each of the built heritage's peculiar and significant aspects.

Not less important is the direct communication that this BIM prototype can obtain through simple graphic interface applications (see Figures 12.11 and 12.12). Today the separation between the different categories within a project has almost completely disappeared, to the point that the designer has to think about how the project can be communicative and stimulating for any user of the building process, starting from the owner and ending up at the last user, which can be a tourist. In order to be able to interact with organisations and users who do not have BIM skills, simplicity and clarity of communication are the cornerstones of a BIM project.

In the same way and with the same objectives, the University of Alicante, through the AEDIFICATIO group, has carried out a parametric diagnostic study of an ancient villa, the Finca Roca, in L'Alfàs del Pi's community, within a project of a more wide-ranging, urbanistic and social scope. Despite the fact that the work of cataloguing and pathological diagnostic study have already been carried out in the CAD area, the coexistence and therefore the completeness of the BIM model (see Figure 12.13) is considered an added value in order to obtain

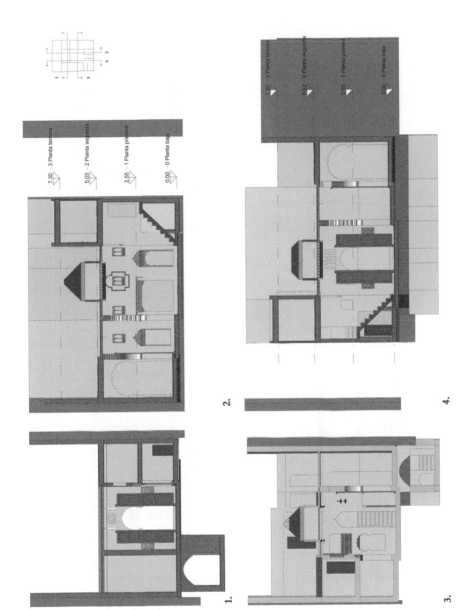

Figure 12.6 BIM modelling.

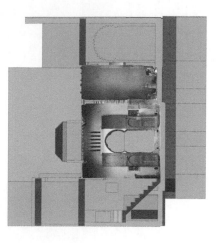
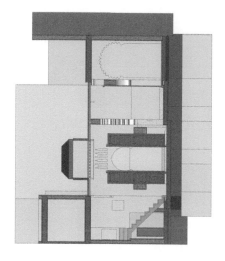
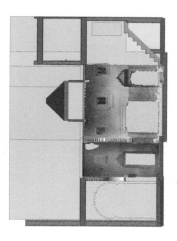
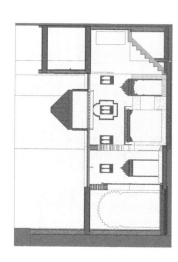

Figure 12.7 Tetuán, pathological analysis.

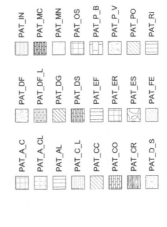

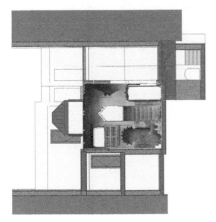

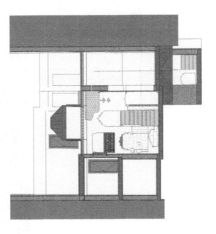

PAT_A_C		PAT_DF		PAT_IN
PAT_A_CL		PAT_DF_L		PAT_MC
PAT_AL		PAT_DG		PAT_MN
PAT_C_L		PAT_DS		PAT_OS
PAT_CC		PAT_EF		PAT_P_B
PAT_CO		PAT_ER		PAT_P_V
PAT_CR		PAT_ES		PAT_PO
PAT_D_S		PAT_FE		PAT_RI

Figure 12.8 Tetuán, pathological analysis.

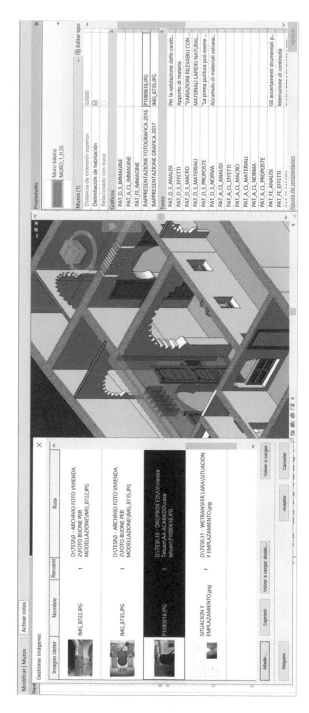

Figure 12.9 Parameterisation of the pathological diagnostic data in the single element.

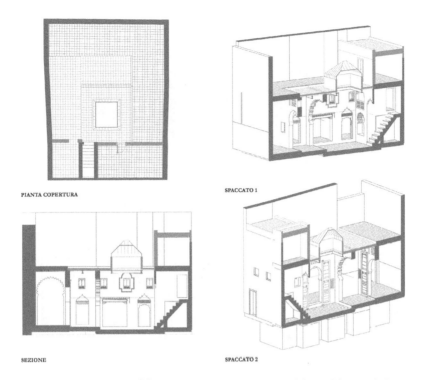

PIANTA COPERTURA

SPACCATO 1

SEZIONE

SPACCATO 2

Figure 12.10 Management of the different temporal phases of the architectural elements.

an optimal diagnostic parametric catalogue of the site. In fact, the possibility of updating the BIM model over time, adding or eliminating information in relation to the actual status of the building, is a significant added value in a restoration project that does not present a timing schedule to date and that in the future could represent the actual state of works carried out, going from "as damaged" to "as is".

The 3D modelling of the Finca Roca, made in detail for each individual architectural element with the creation of special "families", allows a subsequent diagnostic study of the individual elements, thus allowing one to separate the different areas according to the restoration needs that are found (see Figure 12.14). The Finca Roca has been subjected to the pathological coding used in the study of the Moroccan *vivienda*. The fact that this approach can be applied to any building reinforces the concept that the methodology, when applied in a particular way, is exportable to different case studies and to different parametric models.

The updatability over time of the model allows one to carry out different parallel studies, geometric or pathological, that are opportunely separated temporally or typologically. Moreover, it is possible to hypothesise different restoration interventions in relation to the different phases of the project, without losing important information on other project phases or factual states. As mentioned, the possibility of carrying out a pathological study both graphic (*mapeado*) and data-driven (parameterisation of the diagnostic study within the components and of the model

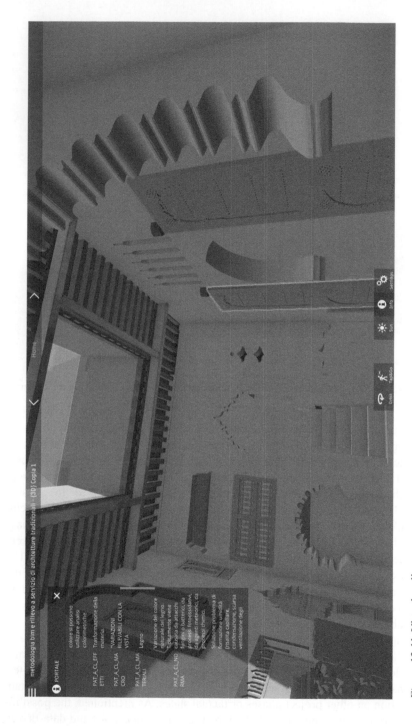

Figure 12.11 Virtual reality.

Figure 12.12 Virtual reality.

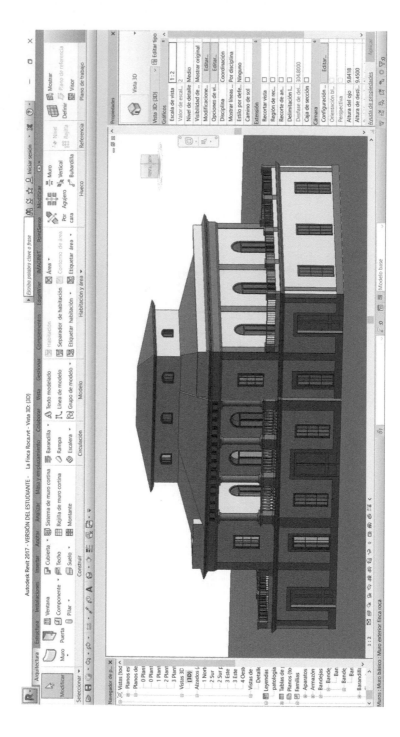

Figure 12.13 3D modelling.

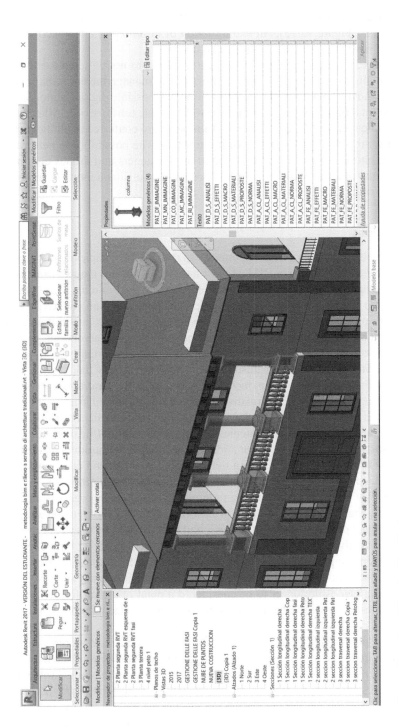

Figure 12.14 Modelling by "families".

families), allows one to understand and control the building with different scales of detail (see Figure 12.15). An extrapolation of information circumscribed and related to what the case requires is also enabled.

Therefore, a project of conservative restoration does not only have as its sole aim the realisation of the project itself, but aims to manage the entire built heritage throughout its whole existence. In addition, the possible digital cataloguing of the different built heritage can be applied to the whole territory of L'Alfàs del Pi. Thus, a uniqueness and a correlation methodology are conferred to all the historical and architectural built heritage of the territory. The possibility to perform an excellent photogrammetric survey through laser scanners and drones eliminates the difficulty that has been found in Morocco, thus realising a more accurate and complete study.

Conclusions

In essence, BIM presents a huge opportunity to manage the complexity of building-related processes. The use of BIM on built heritage of various natures can also be useful in acquiring high-quality information related to the building. This obtained information allows for high-accuracy graphical representation, which in turns helps with prevention and planning of interventions as well as the drafting of the restoration project, in particular in the realisation of single interventions.

Owning a preventive digital catalogue of buildings can certainly facilitate the understanding of damage mechanisms, which are often largely linked to the quality of materials and construction techniques. For example, natural disasters such as earthquakes or simply structural disruptions could be avoided in a layered structure such as the Medina of Tetuán. Moreover, the quality of the BIM that meets the most careful conservation strategies lies in the possibility (the opportunity) to add, in successive moments, information on the state of the building's conservation. This quality allows monitoring over time all changes and alterations, thus having a feedback that consciously orientates the intervention proposals.

In this way, we want to undertake a programmatic and not emergency-like path, considering the built heritage not only in terms of emergency recovery but above all to care for and protect it. As Cesare Brandi had hypothesised in *Teoria del restauro* (1963), preventive restoration is often more important than any other form of protection, as it aims to prevent emergencies, which typically don't allow for a complete rescue of the work of art after the fact. Therefore, the realisation of a "passport" of the built heritage is no longer to be considered a point of arrival for the process; rather, it must be considered a starting point for management and protection.

> Knowledge, rather than as a sum of data, becomes an ethical attitude. An ethical attitude is defined by qualities of humility and ability to listen as well as will-power and willingness to open to new horizons for the future, rather than as a desire to close horizons. The construction of the future can not, in fact, take place starting from the destruction of the past but it is increasingly useful to add, rather than subtract resources from our living environment, with the aim of a more sustainable future.[6]

(Musso, 2016)

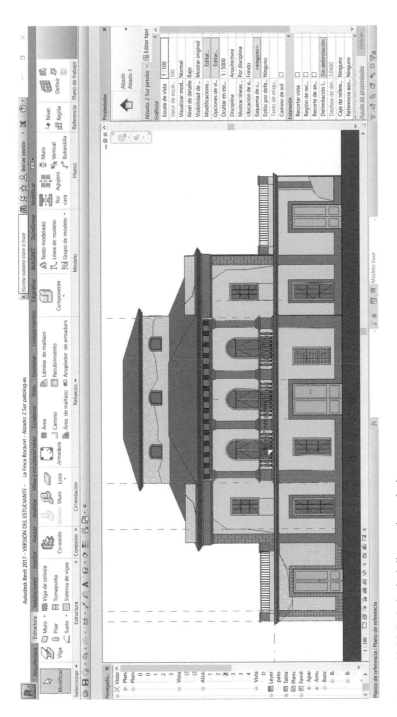

Figure 12.15 Pathological diagnostic analysis.

Notes

1 Translated by the authors from the original in Italian.
2 Translated by the authors from Italian.
3 Translated by the authors from Italian.
4 LODs, divided into both level of development and level of detail, are measures of the amount of information that a certain element model has. According to Building Smart there are five classes of LOD, which contribute to give an estimate of the development design with which the model will be built. The goal is therefore to regulate information in relation to needs. In the case of built heritage and restoration projects we tend to develop the model with a very high LOD, due to the need to describe particular and unique elements both from the geometric point of view and from the information point of view, with the order of size in centimeters.
5 IFC, Industry Foundation Classes, is the format that allows interoperability between BIM platforms and software belonging to different manufacturers, allowing us to export and to subsequently visualise the virtual prototype preserving the parametric properties.
6 Translated by the authors from the original in Italian.

References

Brandi, C. (1963). *Teoria del restauro*. Torino: Einaudi.
Di Giuda, G. M., Maltese, S., Re Cecconi, F., & Villa, V. (2017). *Il BIM per la gestione dei patrimoni immobiliari: Linee guida, livelli di dettaglio informativo grafico (LOD) e alfanumerico (LOI)*. Milano: Hoepli Editore.
Eastman, C., Teicholz, P., Sacks, R., & Liston, K. (2016). *Il BIM. Guida completa al Building Information Modeling per committenti, architetti, ingegneri, gestori immobiliari e imprese*. Milano: Hoepli (or. ed. (2011). *BIM handbook* (2nd Ed.)). New York, NY: John Wiley & Sons.
Grabowski, R. (2010). *CAD & BIM – is there a free pass?* [Web log post]. Retrieved from www.graphisoft.com/ftp/marketing/white_papers/GRAPHISOFT_White_Paper_CADandBIM.pdf
Jaff, M. (2005). *Rilievo fotogrammetrico dell'architettura*. Firenze: Alinea.
Líndez Vílchez, B. (2014). *Tetuán, herencia viva*. Granada: Editorial Universidad de Granada.
Monti, C., & Selvini, A. (2015). *Topografia, fotogrammetria e rappresentazione all'inizio del ventunesimo secolo: strumenti e modalità operative*. Santarcangelo di Romagna: Maggioli.
Musso, S. F. (2016). *Recupero e restauro degli edifici storici: guida pratica al rilievo e alla diagnostica*. Roma: EPC Libri.
New Zealand BIM handbook. (2014). Retrieved from www.mbie.govt.nz/about/whats-happening/news/document-image-library/nz-bim-handbook.pdf
Peirce, C. S. (1935). *Collected papers of Charles Sanders Peirce: Pragmaticism and pragnoaticism, scientific metaphysics*. Cambridge: Belknap Press.

13 The use of IoT and AI to achieve water efficiency in urban environments

Lorena Parra, Albert Rego, Joan Sapena Femenía, Jaime Lloret

Introduction

Now is a time when we are rethinking education, rethinking our cities and are constantly on the lookout for ways in which people can benefit from new technologies. Technology tends to lead to a new kind of wealth for the whole of society, which is distributed in different ways over time. Those responsible for the administration of our cities must think about how to invest in technological progress and take an interest in directing the technological changes of our future cities. Technological change is important because it represents a redistributive paradigm (Atkinson, 2015).

Today, we are in the middle of the third industrial revolution: a revolution of information. Information is owned and huge quantities of it are being sifted within seconds. We can visualise an impending fourth industrial revolution with advances in artificial intelligence (AI) (Gubbi, Buyya, Marusic, & Palaniswami, 2013) and the Internet of Things (IoT) (Atzori, Iera, & Morabito, 2010) already leading to new modes of human organisation. British researcher K. Ashton first used the term the "Internet of Things" (Ashton, 2009). Ashton's idea of IoT is far from other definitions. Ashton values the role of human beings in these technological processes and the role of machines in capturing data about things in the real world. It was estimated in 2017 that IoT was going to be the first source of information for Big Data, even surpassing social networks.

Big Data (Roy et al., 2017) and IoT, applied to the natural environment, are powerful informational sources. According to Mundial (2012) the data are an economic asset comparable to gold currencies. The Garden City was a method of urban planning, it was an early attempt at self-management and it created an intimate relationship between urban and rural areas. The Garden City also represented a paradigmatic model city with both theoretical and practical aspects (ZeXin, Bo, & LiLi, 2011). The year 2007 represented a turning point when cities accommodated more than half of the world's population for the first time. Fifty-four percent of the world's current population lives in urban areas and according to data from a UNO report, that figure will reach 66% by 2050 (News About World Urbanization Prospect, 2014).

G. Lenski discovered that the level of inequality in a society is related to the level of technology (Lenski, 2013). That is why we are interested in the development of AI tools driven by cognitive computing to mitigate our prejudices in decision-making and to promote greater fairness and efficiency in the use of water in urban public spaces. Today we begin to equip our cities with genuine sensory and nervous systems at affordable prices, helping to make them more efficient in terms of resource management, mobility, energy consumption and pollution. We are facing a new relationship between the human being and the environment as a result of technological advances. Rethinking the city as an intelligent organism should enable us to break the deadlock that urban planning has reached. Nevertheless, not only is technology an important factor, but so are creativity and those individuals interpreting said technology. The creativity expressed in any discipline establishes affective bonds between people and places. Creativity provides meaning to our lives and social identity; we think that creativity itself is a necessary quality in the definition of the intelligent city. Nowadays, we cannot understand sustainable development without looking at the interaction between the economic, environmental and social environment where citizen participation is relevant to the success of the processes.

With new technologies and AI we can see things in our environment that we could not see before and we can remember things that we could not remember before. In addition, we can now make calculations and observations about things beyond our vision and beyond our hearing. Applying AI to irrigation, for example, will increase the efficiency of the system and will improve the quality of life in public spaces. We can also see AI as having a broader definition, helping us know ourselves more fully by automating activities associated with human thinking, decision-making, problem-solving, learning, perception and reasoning in many areas of life. Moreover, AI will help us improve our biology and the natural environment, learn more about our limitations and discover new ways of interaction (Ibrahim & Morcos, 2002).

The main target of our proposal is to progress towards sustainable development, especially in the processes of planning and development within local agendas. Using accurate data measurement to distribute resources is also more democratic. The objective is to reach water supply sustainability through different types of sensors in the green areas of cities and homes, where water smart meters will monitor water usage. The IoT research on irrigation that we are promoting will lead to qualitative improvements in the technical sense, including greater efficiency, productivity and reliability. Although the most remarkable applications of AI are currently taking place in genetics, we consider using sensor data through AI to predict the best times to irrigate and achieve water sustainability in urban environments, thus responding to the demands of sustainability (Madurga, 2005).

The rest of the chapter is structured as follows: Section 2 shows the main proposals and research in the fields of IoT and AI. The proposal is described in Section 3. Section 4 presents the requirements to implement the proposed system in different scenarios. Finally, conclusions and future work are discussed in Section 6.

Related work

In this section, we present the state of art on IoT and AI solutions for smart irrigation. IoT and AI (Mali & Kanwade, 2016) offer promising operations and ways of interacting with the environment through autonomous irrigation. The irrigated area has not only a technological identity, but reflects above all a social choice and will be the result of social decision-making. For that reason, any formal and informal work related to the search for hydraulic solutions in urban irrigation requires cooperation between different social agents. Solutions based on automatic timers are being replaced by digital technology. This research is of interest for different areas of knowledge, different institutions and also for the maker community within our society.

Related works have investigated the factors that affect the present and future demand for water (Seckler, 1998). This line of research looks at the density of vegetation, surface and typology of public and private gardens, because those constitute the elements that most affect water consumption in urban areas (Morote Seguido, 2016). Important steps are being taken in the recovery and protection of resources in the areas of water management, energy, recycling and waste recovery.

Ubiquitous technologies (Specht, Tabuenca, & Ternier, 2015) and IoT are identified in different articles as trends that are opening up new avenues for the development of new technologies applied to the green areas in cities. We can see that technology has migrated from laboratories, classrooms, factories and warehouses to the public spaces of cities, homes and fields. This technological migration has been possible thanks to digital developments that allow mobility in different devices and interconnection between devices. Currently, most irrigation controls can be configured by hand from mobile apps or smartphones. In this chapter, a few types of automated controls are analysed: works on irrigation systems, detection of alarms, irrigation in real time and smart watering to analyse the level of vegetation or other environmental data.

Gutiérrez, Villa-Medina, Nieto-Garibay, and Porta-Gándara (2014) showed improvements in the irrigation system. Their aim was to show a programmed irrigation system powered by photovoltaic panels. Their system was tested in a sage crop field for 136 days. Water savings were up to 90% compared to water consumption in the agricultural areas with traditional irrigation. Another innovative hybrid system for smart agriculture is introduced. It is called AgroTick and it was developed by S. Roy et al. AgroTick is an IoT based system supported with mobile interface and designed using technology modules like cloud computing, embedded firmware, a hardware unit and Big Data analytics (Roy et al., 2017).

Secondly, the iRain system ("intelligent rain"), developed by Caetano, Pitarma, and Reis (2014) aims to supply autonomous irrigation and provide efficiency in urban gardens. iRain consists of a net of sensors and actuators with Zigbee communication and control software for storage. It is able to monitor data on a web portal in real time. The system also gives preference to irrigation at night and during periods of light wind. The purpose of the proposed system is to obtain more accurate and efficient irrigation through weather forecasting and monitoring

environmental conditions and soil moisture. This procedure can be manually activated for each garden from the developed web portal.

Finally, we could mention smart metering, which gives us a deep knowledge
about the use of water in cities, a subject to which not enough attention has been
paid so far (March, Morote, Rico, & Saurí, 2017). Therefore, we propose an
integral work that uses climatic, humidity and conduction data to determine the
best moments for irrigation as well as household water consumption. Techniques
like ours are not used as often in gardening and our aim is to use a considerable
amount of data and a smart algorithm within a software-defined network (SDN)
that gives scalability and security in water consumption.

Proposal description

Purpose

The proposed system has been designed to enhance the efficiency of water usage
in smart cities. This efficiency is achieved in two different ways:

> The first one is to adjust the irrigation needs in order to avoid water waste.
> The adjustment will be conducted by controlling the exact amount of water
> needed. Commonly, the irrigation process is executed according to pre-set
> values based on the grass type and the season of the year. However, climatic
> conditions, topology, soil type or shadow conditions are not considered.
> Those conditions cause that in different parts of the same garden the water
> demand is different. Thus, irrigation requirements change from one part of
> the garden to the other. For this reason, it is necessary to consider the humid-
> ity in the soil before distributing water in the gardens. This methodology is
> widely used for agricultural purposes and it is part of the precision agriculture
> method. Nevertheless, these techniques are not used as often in gardening.
> We can find some examples in literature where authors employ sensors to
> adjust the irrigation in crops.

We can find some examples in literature where authors employ sensors to adjust
the irrigation in crops as the proposal from Marín et al. (2017) or the proposal
from Parra et al. (2018).

> The second one is to define a series of rules to ensure that the irrigation is
> achieved in the best moment. The most important rule is to water the gardens
> only during nighttime. Watering only at nighttime will reduce the fast evapo-
> ration of the water which can cause many problems. Moreover, the system
> considers the water consumption of houses and business in order to decide
> the best moment to irrigate. The objective is to find the best moment to use
> water for irrigation; the time usually presents itself when houses and business
> are not using water. Thus, the pressure over the water supply sources will be
> reduced. It is better to have similar water consumptions along the 24 hours of
> the day when there are different demand peaks.

For that purpose, our system must be able to read data on the state of the gardens and on the water consumption carried out by houses and businesses. In addition to reading a high amount of information, integration and pattern identification and prediction are crucial. Therefore, a smart system able to take decisions based on a set of given rules is needed. Moreover, the proposed system needs to ensure the security of the data sent from houses and businesses and it must be easily scalable due to continued city growth.

Urban scenarios

In this section, we are going to identify and describe the urban scenarios where our proposal can be developed. First, the different areas in the cities are defined. Then, the employed sensors in each area are shown. For our proposal, we are going to differentiate two separate urban areas. Residential areas, where homes and business are located, usually consume high amounts of water during the day and the first hours of the night. Water is mainly consumed for domestic purposes such as cleaning, cooking and hygiene, among others. In small percentages, some houses have their own gardens and use water for irrigation purposes or to fill swimming pools. Water is used in business for almost the same purposes as at home. Depending on the business, it can use more or less water and maximum consumption can occur at different hours.

The second area to analyse as far as water consumption is the green area. In green areas water is mainly consumed for irrigation purposes, but it also can be used in drinking water fountains or decorative fountains. The biggest water consumption occurs during the night, when the gardens are irrigated. According to the climatic conditions, grass requirements and soil type, the irrigation needs may change. For irrigation purposes, we can distinguish between walkable green areas and non-walkable green areas. On one hand, walkable green areas are considered all the green areas where people's presence is common, like parks, gardens, sport areas or recreational areas. On the other hand, non-walkable green areas are all the green areas where people's presence is not expected, like roundabouts and similar areas. Non-walkable green areas can be watered anytime in the night. In walkable green areas, we can distinguish two categories; the first one is areas with opening and closing hours like sport areas. Those areas can be watered in the night during the closing period. The second one is green areas without opening and closing times. Those must be watered during the night, but ensuring that nobody is there.

Thus, we have four different types of areas in the city: (1) residential areas, (2) non-walkable green areas, (3) walkable green areas with no schedule, and (4) walkable green areas with a schedule.

Employed sensors

In residential areas, sensors will be deployed to measure water consumption. Those sensors are integrated into the water supply system and they are called smart meters. Smart meters are digital electronic devices that can collect data on supply use and send it to the utility (Lloret, Tomás Gironés, Canovas Solbes, &

Parra-Boronat, 2016). The supplies can include water, electricity or gas. In this case, the smart meters will monitor water use.

In green areas, sensors to measure the garden conditions and actuators that activate/stop the irrigation system will be deployed. The sensors that control the garden status include: (1) rain (Figure 13.1), (2) soil moisture (Figure 13.2), temperature (Figure 13.3), wind (Sendra, Parra, Lloret, & Jiménez, 2015) and solar

Figure 13.1 Rain sensor.

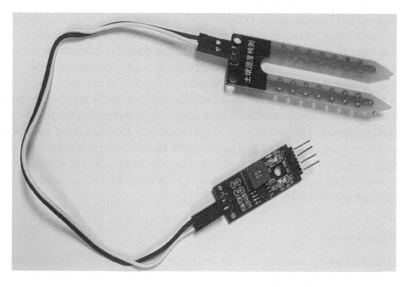

Figure 13.2 Soil moisture sensor.

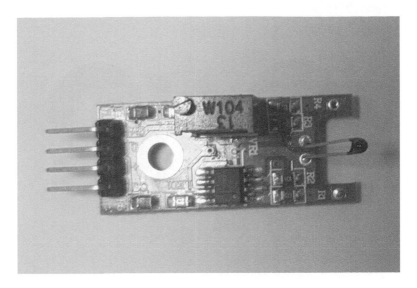

Figure 13.3 Temperature sensor.

Figure 13.4 Solar radiation sensor.

radiation sensors (Figure 13.4), which are located in atmospheric stations placed in the gardens, (3) water flow sensors in the water pipes, and (4) presence sensors (Figure 13.5). Soil moisture sensors define the amount of water needed by a portion of the garden. Temperature, rain, wind and solar radiation sensors are used to define the best moment to irrigate. The current sensors in pipes are employed to

Figure 13.5 Presence sensor.

ensure that there is no broken pipe. Presence sensors, which are placed along the gardens without a special schedule, define, in conjunction with climate sensors, the best moment to irrigate.

In Table 13.1, we show the information related to the data gathered and sent by each sensor. In this table, we include the maximum and minimum possible values. The sensors may gather values higher than the maximum and lower than the minimum, but those values are not expected in the deployed areas. Moreover, the table includes the sensitivity required from the sensor. The sensors could have higher sensitivity, but for our application we require less sensitivity. The objective is to minimise the bits used to send data from each sensed parameter.

Architecture

The architecture proposed is a combination of nodes that measure data and a network infrastructure. This infrastructure not only has to be able to transmit the data but also has to provide a way to act intelligently and efficiently. Therefore, the network architecture chosen is SDN. That network builds the core of the system in terms of communication. It also can be connected to the Internet to use external services.

In order to improve the simplicity of the figures, the actuators that can appear together with the sensors in their networks are omitted. When a sensor appears in the figure, it can be both a sensor and an actuator. Figure 13.6 shows all the different actors in the architecture and their connections. On the one hand, the data-gathering actors are connected to the core network. These actors can be either smart meters or

Table 13.1 Possible values gathered by each sensor

Parameter	Min.	Max.	Interval	Sensitivity	Possible values	Bits
Presence (yes/no)	0	1	2	1	1	1
Moisture (%)	1	100	100	3.5	28.6	32
Water (yes/no)	0	1	2	1	1	1
Temperature (°C)	−19	40	60	2	30	32
Rain (yes/no)	0	1	2	1	1	1
Wind (km/h)	0	31	32	2	16	16
Light (lux)	0	99.999	100	2000	50	64

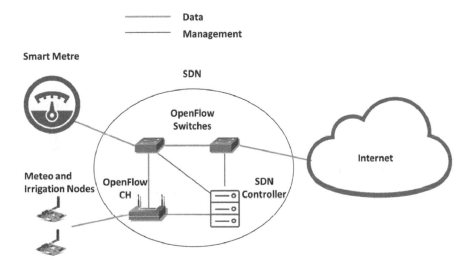

Figure 13.6 Architecture of the proposed system.

sensor nodes. Furthermore, the sensor nodes can be meteo nodes (weather stations) or irrigation nodes. The core network is an SDN composed by OpenFlow-enabled switches and access points (AP). On the other hand, the SDN can be also connected to the Internet. That is the current data path of the system. However, an SDN controller is needed to manage the SDN and it is connected to every node in the SDN. In the figure, the connections used for managing the SDN are painted brown.

This architecture is deployed in the Smart City and it creates several data source networks. These networks can be composed either by a smart meter or by a wireless sensor network (WSN). A WSN is made up of sensors gathering different data such as humidity and temperature. In addition, there are some actuators like a garden hose for irrigation. Depending on the kind of source network, the gateway that interconnects the SDN with that source network varies. If the source network is a smart meter, the gateway will be an OpenFlow switch. Otherwise, WSNs need an access point, because they are wireless. The access point is the cluster head

(CH) of the WSN as well. These two different source networks are displayed in Figure 13.7, where an example of multiple networks connected in the Smart City using the architecture is presented.

Finally, the layered architecture is shown in Figure 13.8. The nodes in the architecture are classified into different categories or layers depending on their function. The first layer contains the nodes whose goal is to gather the data. The data is propagated through the network nodes, which belong to the network layer. The

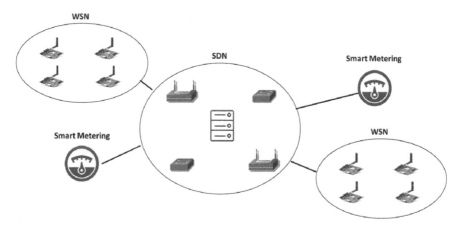

Figure 13.7 Example of multiple networks connected in the Smart City.

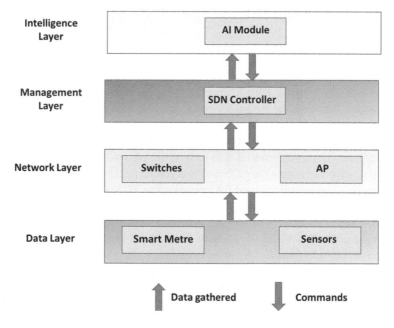

Figure 13.8 Layer architecture.

data gathered by the smart meters and the sensors are sent to the SDN controller with the network statistics gathered by the network nodes. The controller provides management functions for the data and for the network, being able to command new measurements or data sending if the data is not well received. Furthermore, the controller is able to change network paths or processes if the network conditions change drastically. Nevertheless, the decisions needed to be taken are determined by the intelligence layer. The AI module, a software module that resides on the SDN controller, executes rules to decide what to do in attending to the data gathered. The decisions are transformed into commands that are propagated down-way in the layered architecture. The actuators will execute the decided action, for example, save water or irrigate. The sensors and the smart meters can act too, changing the measurement frequency or executing other functions.

Messages

Once the architecture and the aim of the system have been detailed, the next step is to define the messages used to provide the functionality to the system. The OpenFlow standard is used by the controller for gathering data from the network nodes. For example, when a sensor sends data to the CH, an OpenFlow message is used to forward the packet to the controller through the secure channel that connects the controller to the network nodes. However, some custom messages are needed to implement all the functionality of the system. Moreover, besides the messages used to communicate the data gathered by the sensors, other messages are needed to implement functions. The custom messages defined to be used by the system are the following:

- ACK message: used to inform that the procedure carried out has been properly executed.
- Data: used to communicate the data collected from the sensors to the CH. This message is sent from any sensor or smart meter to the CH.
- Data Request: used by the SDN controller to request data to the sensors in a specific time instant. The sensors will perform a new measurement and will respond with a data message.
- Sleep: used to put a specific sensor or actuator to sleep during a specific period of time. It is sent from the SDN controller to any sensor or smart meter.
- Activation message: used to wake a sleeping sensor or actuator immediately. The SDN sends this message to a sleeping sensor or actuator.
- Act time message: used to specify the waking time of a sleeping sensor or actuator. Like the activation message, it is sent from the SDN controller to a sleeping sensor or actuator.
- Stop irrigation message: used by the sensors to make the irrigation stop. This message is sent from the sensor to the actuators.
- Error message: used to inform that some error has appeared during the procedure.

All these messages are sent in encapsulated form in User Datagram Protocol (UDP). They include a header where the type of message and the length of the

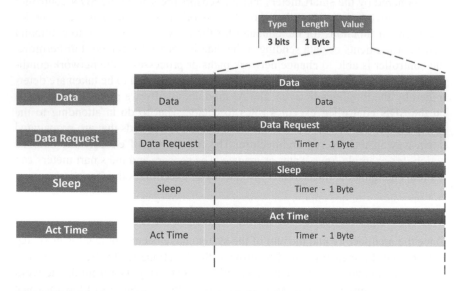

Figure 13.9 Messages with non-empty value field and their structure and length.

payload are detailed. Then, the data is presented. This structure is displayed in Figure 13.9, where the messages are presented. Not all eight messages appear in the figure, because some of them are empty messages. They are only composed by the type and length header, while the length header value is zero. Those messages are ACK message, activation message, stop irrigation message and error message. All of them are direct commands or info messages. The other messages need some fields in the value header to communicate data. Their fields are detailed in Figure 13.9 among the type, length and value headers. Accordingly, the system is constantly using OpenFlow standard messages to collect statistics from the SDN nodes and to forward messages to the controller. In addition to the previously defined messages, the system executes actions and communicates the data from the sensors and smart meters.

The messages are sent through a UDP datagram. Therefore, the total size of the packets, either sent from or received by the sensors or actuators, is 79 bytes from the datagram, 3 bits from the type header, 1 byte from the length header and the length of the value header (1).

$$Size = Size_{UDP} + Size_{Headers} \tag{1}$$

Where $Size_{UDP} = 79\,Bytes \tag{2}$

$$Size_{Headers} = Size_{Type} + Size_{Length} + Size_{Value} \tag{3}$$

$$Size_{Type} = 4\,bits \tag{4}$$

$$Size_{Length} = 1 \ byte \qquad (5)$$

and $\quad Size_{Value} = Length \qquad (6)$

The length of the packet sent depends on the type of packet. The ones that do not appear in Figure 13.9 will have a size of 643 bits. However, data request message, sleep message and act time message will have a size of 651 bits. The size of the data message will depend on the type of node. We manage four different kinds of nodes, described in Table 13.2, where the data sent by each kind of node is detailed. Each node sends a different amount of data; they are specified in the system just as in Table 13.2. Therefore, the system is able to identify which kind of data it is receiving and how many bits it should read depending on the length header. This prevented us from using a "node type" or "data type" field in the data message, thus reducing the message size and energy consumption. The data message size will depend on the node; it is 757 bits for a weather station node, 677 bits for an irrigation station node and 647 bits for a smart meter. In Table 13.3, each kind of data type gathered by a different sensor is defined and its length is detailed. With the messages presented, the communication process and the size of the messages have been discussed. The next point to discuss is the system's algorithm.

Algorithm

The irrigation process consists of a periodic evaluation of the scenario. Usually, every hour, the sensors and smart meters send data to the SDN controller. The

Table 13.2 Data quantity and type sent by each kind of node

Node type	Data sent by the node	Data length
Weather station node	Temperature, rainfall, wind, light	114 bits
Irrigation station node	Humidity, water flow	34 bits
Smart meter	Consumption	4 bits

Table 13.3 Data length of each data type

Data type	Length
Temperature	32 bits
Rainfall	2 bits
Wind	16 bits
Light	64 bits
Humidity	32 bits
Water flow	2 bits
Presence	2 bits
Consumption	4 bits

algorithm then evaluates the conditions for irrigating. The following factors need to be taken into account before irrigating:

- If it is raining, the irrigation cannot start. The system should wait until the rain stops in future iterations.
- If the forecasting service indicates that it will rain at the irrigation time, the irrigation is postponed.
- If there are people in the irrigation area (in those where people can walk on the grass), the irrigation does not start.
- If there are gusts of wind greater than 20 km/h, the irrigation is cancelled.
- If the forecasting service indicates that there will be a freeze, the irrigation is cancelled.
- If the forecasting service indicates that there will be gusts of wind, the irrigation is immediately started.
- If none of these conditions are true, irrigation will start only if it is the scheduled time. This time is a period between the consumption peaks of the households at night and in the morning. The irrigation must start after the night peak and before the morning peak. Furthermore, if the area has an opening and closing time, the irrigation will be performed only after the closing time and before the opening time.

The flowchart of that irrigation process is shown in Figure 13.10. First, all the data needed to be read is obtained. The systems use data from different sources

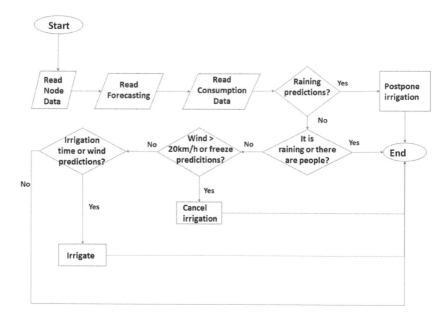

Figure 13.10 Algorithm of the irrigation process.

including sensor data, either obtained each hour from the sensor or consulted by the controller, the forecasting consulted in the web service and the consumption data gathered from the smart meters. After having gathered all this data, the conditions are evaluated and the decision is made. This process is repeated every day several times until the irrigation is performed.

Some details of the process must be discussed. On the one hand, irrigation will stop when the humidity sensor measures a humidity value greater than a defined threshold. This process is performed by using the "stop irrigation" message defined previously. On the other hand, the data of the consumption is processed by the AI module. Using a statistic model, the future peaks of consumption are predicted. Moreover, changes in this consumption habit are also predicted. For example, if the consumption peak is delayed from 9:00 p.m. to 10:00 p.m. in summer, the system will adapt to that change. Therefore, the irrigation period will change dynamically, depending on the users, and water consumption problems during the irrigation process will be avoided. However, prediction of consumption periods is not the only thing that the AI module can provide to the system. Some issues and problems, like leaks or frauds, can also be detected by analysing the consumption histogram.

Requirements for the implementation of the proposal

In this section, the network requirements as if we were to implement the aforementioned proposal in urban environments are presented. We are going to show the use of the proposal in different types of cities. First, we are going to describe five different scenarios.

* The first scenario is a small village with 100 houses, 3 businesses and a few green areas > 1000 m². Its green areas are mainly composed of roundabouts and small plots with grass and trees.
* The second scenario corresponds to a mid-size town with 5,000 houses, 100 businesses and 20,000 m² of green areas. The green areas are composed of roundabouts, municipal gardens and a sport area.
* The third scenario is another mid-size town with 5,000 houses, 100 businesses as the previous case, but with bigger green areas. A total of 700,000 m² of green area are spread around town. In addition to the aforementioned green areas in this town, there is a golf course.
* The fourth scenario is a city with 200,000 houses, 25,000 businesses and 500,000 m² of green areas.
* The fifth example is a big city with 1,500,000 houses, 400,000 businesses and 62,000,000 m² of green areas. The green areas include different sport areas, roundabouts, big gardens and recreational areas.

Now, we are going to show the requirements if we pretend to deploy the proposed system in those scenarios. Thus, we are going to make some assumptions. The first is to assume that in all the houses and businesses there is a smart meter installed. Next, the meteorological stations, where the temperature, rain, wind and solar radiation sensors are located, are placed in the green areas each 10,000 m^2. In the green areas, the humidity and the current sensors are placed at each sprinkler. This placement is called the irrigation control station. We can assume that each sprinkler covers an area of approximately 37m^2. The sprinklers have an overlapping of 25%. Thus, we can easily calculate the number of smart meters and stations located in each one of the aforementioned examples (Table 13.4).

Once we set the number of smart meters, meteorological stations and irrigation control stations, we can calculate the amount of traffic generated in each case. The meteorological stations send messages of 114 bits of data and 643 bits of headers. The irrigation stations send messages of 34 bits of data and 643 bits of headers. Finally, the smart meters send messages of 2 bits of data and 643 bits of headers. Thus, we can calculate the generated traffic when the SDN request data from all the nodes in each one of the aforementioned examples. The data obtained can be seen in Figure 13.11. In the example with the smallest

Table 13.4 Number of nodes for different purposes in each scenario

City	Smart meters	Meteorological stations	Irrigation control stations
1	103	1	28
2	5,100	2	541
3	5,100	70	18,919
4	225,000	50	13,514
5	1,900,000	6,200	1,675,676

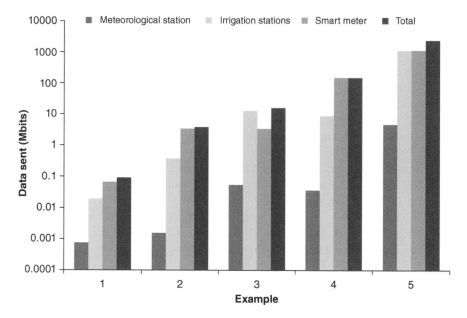

Figure 13.11 Data sent in each scenario to the SDN.

town the data sent are 86.35 Kbits and, in the biggest city, the data are 2368.43 Mbits. We can see that the lowest traffic comes from the meteorological stations in all the cases. In four cases the major traffic component is related to the smart meters. Only in the third example, when a big surface is covered by green area, the data generated by the irrigation stations are the major part of the generated traffic.

Conclusions

In this work, smart irrigation systems for urban lawns has been shown. The system uses data from different sensors placed in homes, gardens and water tubes. The objective of this work is to create a smart system able to decide the best moment to start irrigation according to a series of given rules. Smart meters are used to monitor home water demand in order to start irrigating when the system detects that there is no water demand in the dwelling. Several meteorological stations can be used to monitor the climatic conditions in order to delay the irrigation if it is raining or there is too much wind or too low temperatures. In other words, we can save water if we wait to irrigate if it is raining, thus avoiding misuse of water, or avoiding damages in case of possible frost. Moreover, the system has sensors placed inside the water tubes. These sensors are monitoring water flow and humidity in the sprinklers. The sensors in the tubes are placed to detect breakages and the soil moisture sensors close to the sprinklers monitor the water presence in the soil and determine when it is necessary to irrigate and when it is necessary to stop it. Thanks to the use of data and smart algorithms, the system expects to save some of the water used for irrigation and to reduce the water demand peaks. We have demonstrated the network bandwidth required in different scenarios when this system is used. SDN is used in this system to ensure the future scalability, data security, and the inclusion of the smart algorithm in the network.

As future work, we hope to include this type of system in other environments where water demand must be monitored in rural areas. In addition, we would like to use similar systems in the Smart City to improve the sustainability of other issues, such as public illumination.

References

Ashton, K. (2009). That "internet of things" thing. *RFID Journal, 22*(7), 97–114.

Atkinson, A. B. (2015). *Inequality*. Cambridge: Harvard University Press.

Atzori, L., Iera, A., & Morabito, G. (2010). The internet of things: A survey. *Computer Networks, 54*(15), 2787–2805.

Caetano, F., Pitarma, R., & Reis, P. (2014). *Intelligent management of urban garden irrigation*. Proceedings of the 9th Iberian Conference on Information Systems and Technologies (CISTI, Barcelona) (pp. 1–6). Piscataway, NJ: IEEE Press.

Gubbi, J., Buyya, R., Marusic, S., & Palaniswami, M. (2013). Internet of Things (IoT): A vision, architectural elements, and future directions. *Future Generation Computer Systems, 29*(7), 1645–1660.

Gutiérrez, J., Villa-Medina, J. F., Nieto-Garibay, A., & Porta-Gándara, M. Á. (2014). Automated irrigation system using a wireless sensor network and GPRS module. *IEEE Transactions on Instrumentation and Measurement, 63*(1), 166–176.

Ibrahim, W. A., & Morcos, M. M. (2002). Artificial intelligence and advanced mathematical tools for power quality applications: A survey. *IEEE Transactions on Power Delivery, 17*(2), 668–673.

Lenski, G. E. (2013). *Power and privilege: A theory of social stratification.* Chapel Hill, NC: UNC Press Books.

Lloret, J., Tomás Gironés, J., Canovas Solbes, A., & Parra-Boronat, L. (2016). An integrated IoT architecture for smart metering. *IEEE Communications Magazine, 54*(12), 50–57.

Madurga, M. R. L. (2005). Los colores del agua, el agua virtual y los conflictos hídricos. *Revista de la Real Academia de Ciencias Exactas, Fisicas y Naturales de Madrid, 99,* 369–389.

Mali, N., & Kanwade, P. A. (2016). A review on smart city through Internet of Things (IOT). *International Journal of Advanced Research in Science Management and Technology, 2*(6).

March, H., Morote, Á. F., Rico, A. M., & Saurí, D. (2017). Household smart water metering in Spain: Insights from the experience of remote meter reading in Alicante. *Sustainability, 9*(4), 582.

Marín, J., Rocher, J., Parra, L., Sendra, S., Lloret, J., & Mauri, P. V. (2017). *Autonomous WSN for lawns monitoring in smart cities.* 2017 IEEE/ACS 14th International Conference on Computer Systems and Applications (AICCSA), Hammamet, Tunisia (pp. 501–508). New York, NY: IEEE Press.

Morote Seguido, Á. F. (2016). El uso del agua en los jardines de las urbanizaciones del litoral de Alicante. Prácticas de ahorro y sus causas. *Investigaciones Geográficas (Esp),* (65), 1–18.

Mundial, F. E. (2012). *Big data, big impact: New possibilities for international development. Foro Económico Mundial.* Cologny, Suiza. Retrieved May 2, 2018, from www3. weforum.org/docs/WEF_TC_MFS_BigDataBigIm-p act_Briefing_2012.pdf

Parra, L., Rocher, J., García, L., Lloret, J., Tomás, J., Romero, O. (. . .) Roig, B. (2018). Design of a WSN for smart irrigation in citrus plots with fault-tolerance and energy-saving algorithms. *Network Protocols and Algorithms, 10*(2), 95–115.

Roy, S., Ray, R., Roy, A., Sinha, S., Mukherjee, G., Pyne, S. (. . .) Hazra, S. (2017). *IoT, big data science & analytics, cloud computing and mobile app based hybrid system for smart agriculture.* IEEE 2017, 8th Annual Industrial Automation and Electromechanical Engineering Conference (IEMECON) (pp. 303–304). New York, NY: IEEE Press.

Seckler, D. W. (1998). World water demand and supply, 1990 to 2025: Scenarios and issues. *Iwmi Research Report, 19,* 1–41.

Sendra, S., Parra, L., Lloret, J., & Jiménez, J. M. (2015). Oceanographic multisensor buoy based on low cost sensors for Posidonia meadows monitoring in Mediterranean sea. *Journal of Sensors,* 1–23.

Specht, M., Tabuenca, B., & Ternier, S. (2015). Tendencias del aprendizaje ubicuo en el internet de las cosas. *Campus Virtuales, 2*(2), 30–44.

United Nations Department of Economic and Social Affairs (DESA) (2014) *World Urbanization Prospects.* Retrieved May 2, 2018, from www.un.org/es/development/desa/news/population/world-urbanization-prospects-2014.html

ZeXin, L., Bo, L., & LiLi, Z. (2011). Water resource's ecological counteraccusation towards garden city. *Procedia Environmental Sciences, 10,* 2518–2524.

Index